Rural Geographies

Rural Geographies provides a critical, contemporary and accessible introduction to rural change by using geographical ideas to understand current issues affecting the countryside.

The book discusses how the countryside has been studied by geographers across a range of different scales, from village community to the global countryside. Each chapter provides a concise and well-illustrated introduction to a key theme in rural geography, using current literature and contemporary examples. The book is divided into four sections that cover rural contexts, changes, contests and cultures. The volume takes a global perspective but is largely centred on the Global North, reflecting the tradition of scholarship in rural geography.

Rural Geographies is driven by thinking in human geography. It reflects how major paradigmatic changes in the discipline have impacted, and have been informed by, the sub-discipline of rural geography. The aim is to introduce key ideas and concepts that will teach students the critical skills necessary to analyse rural issues themselves. The text will be a valuable resource for undergraduate students studying rural geography and rural studies.

Richard Yarwood is Professor of Human Geography and the Director of the Doctoral College at the University of Plymouth, UK.

Rural Geographies

People, Place and the Countryside

Richard Yarwood

Routledge
Taylor & Francis Group
LONDON AND NEW YORK

Cover image: © Getty Images

First published 2023
by Routledge
4 Park Square, Milton Park, Abingdon, Oxon OX14 4RN

and by Routledge
605 Third Avenue, New York, NY 10158

Routledge is an imprint of the Taylor & Francis Group, an informa business

British Library Cataloguing-in-Publication Data
A catalogue record for this book is available from the British Library

Library of Congress Cataloging-in-Publication Data
Names: Yarwood, Richard, author.
Title: Rural geographies : people, place and the countryside / Richard Yarwood.
Description: Abingdon, Oxon ; New York, NY : Routledge, 2023. | Includes bibliographical references and index.
Identifiers: LCCN 2022038532 (print) | LCCN 2022038533 (ebook) | ISBN 9781138327955 (hardback) | ISBN 9781138327993 (paperback) | ISBN 9780429448966 (ebook)
Subjects: LCSH: Rural geography. | Rural conditions. | Rural population. | Human geography.
Classification: LCC GF127 .Y37 2023 (print) | LCC GF127 (ebook) | DDC 307.72—dc23/eng20230106
LC record available at https://lccn.loc.gov/2022038532
LC ebook record available at https://lccn.loc.gov/2022038533

ISBN: 9781138327955 (hbk)
ISBN: 9781138327993 (pbk)
ISBN: 9780429448966 (ebk)

DOI: 10.4324/9780429448966

Typeset in Bembo and Helvetica Neue LT Pro
by Apex CoVantage, LLC

For Ruth, Elizabeth and William.

Contents

Figures

Tables

Boxes

About the Author

Richard Yarwood is Professor of Human Geography and the Director of the Doctoral College at the University of Plymouth, UK. He is an active researcher in rural geography, regularly publishing in the *Journal of Rural Studies* and other geographical journals. His work has examined social and cultural change and has included work on housing, policing, population change, service provision, volunteering and animal geographies. His research is international in its scope and impact, reflecting research projects in the UK, Europe, Australia, the USA and New Zealand. He has published various books on rural issues, as well as on citizenship and policing. Richard has taught rural geography to undergraduates and postgraduates for over 30 years.

Foreword

Change is an important theme in rural geography. I doubt, though, whether any rural geographers could have predicted any of the extraordinary changes that impacted the world and its rural areas during the period in which this book was written.

I started writing this book in 2018. In 2020 Covid-19 was declared a pandemic; at the end of the same year, the UK left the European Union and, in 2022, Russia invaded Ukraine. These momentous events have emphasised the global significance of the countryside. Covid-19 led to restrictions on international movement and concerns that a lack of migrant labour would leave crops unharvested and meat unprocessed. The closure of borders fuelled fears that food would be prevented from reaching supermarket shelves, leading to panic buying that itself caused shortages. In the UK, these issues were compounded by new, post-Brexit border controls that reduced much of Kent to a lorry park. At the time of writing, the war in Ukraine has led to drastic declines in food supplies that have contributed to shortages and inflation. These events have highlighted the global nature of the countryside and the significance of understanding how rural spaces connect with other places across the world.

When this book was reviewed, one referee felt that the pandemic had led to radical changes in rural places that deserve greater scrutiny by geographers; another suggested the pandemic, while profound and significant, would soon pass and the countryside would return to relative normality. Likewise, one reviewer felt that Brexit should have been discussed in detail; another felt that its main significance was confined to the UK and, given an international readership, did not need mentioning in any depth. All were right. These events have drawn attention to many important, everyday rural issues (such as growing and distributing food, leisure activities, environmental change and social inequalities) that are rarely given wider attention. They highlighted the global nature of the countryside and the significance of understanding how rural spaces connect with other places across the world. In short, this is why rural geography matters!

As shops became re-stocked and international travel has resumed, the countryside has returned to the background of public life. Future research will evaluate the extent to which these changes are long lasting or an anachronism. My approach in this book has, therefore, been to acknowledge, where possible, the significance of these changes but, at the same time, maintain a focus of the two main aims of the book.

First, it seeks to outline the ways in which the countryside is changing over time and space. In doing so, it highlights the significance of the countryside and emphasises why it is important to understand it. Second, it seeks to analyse rural change through the lens of human geography, assessing how major paradigmatic changes have informed and have been informed by rural geography. The intention is not to champion a particular approach or the latest idea but, rather, to introduce how different geographical ideas and concepts have led to a wider understanding of the countryside.

There has been a greater appreciation of the global countryside (Woods, 2007) and the ways that rurality connects the North and South. Consequently, more work from and about the Global South is starting to appear in rural geography literature. There is undoubtedly a need to widen the scope of rural geography, but this book tends to focus on the Global North. In part, this is pragmatic – given the limitations of space, it is only possible to cover so much in one volume – but it also reflects that 'rural geography has almost exclusively been concerned with rural spaces, issues, and processes in the Global North' (Milbourne, 2017: 1).

This book would not have been possible without the support and help of many people. I would like to thank the Ruralia Institute of the University of Helsinki Mikkeli, Finland, for hosting me as a visiting scholar in 2018. The Ruralia Institute's Visiting Scholars Programme allowed me some space and time to start the book and gain new perspectives on rural issues.

I am hugely grateful to Jamie Quinn for his work on the figures and illustrations in this book. He's managed to turn my scribbles and vagaries into clear and concise diagrams that do so much to illuminate complex rural processes. Thanks again, Jamie! Any errors are my own. I also appreciate all those people and organisations that have allowed me to reproduce their photos and material – I hope that I have done them justice and brought your work to a wider audience that it deserves.

I'm also extremely grateful to Claire Maloney, Egle Zigaite and Andrew Mould at Routledge for supporting this book, keeping me on track and overseeing its final production. Particular thanks also go to the anonymous referees who reviewed the book proposal and read the draft manuscript: your comments were helpful, critical and, above all, supportive. These, in no small way, gave me the confidence to push on with the book and complete it, particularly when I had my own doubts.

The book reflects my own interests, understanding, positionality and background. In attempting to paint a broad picture of rural geography, I have inevitably skimmed over some areas, especially those that are less familiar to me. My hope is the book provides a broad introduction to rural geography and signposts the paths that students can follow to develop their own knowledge of specific topics.

The book also reflects my own journey through rural geography, and as such, I am grateful to all those colleagues who have supported and influenced me at Aberystwyth University, the University of Worcester and the University of Plymouth. I would also like to acknowledge the support, friendship and scholarship of The Rural Geography Research Group of the Royal Geographical Society, which celebrated its 50th birthday in 2022.

I would also like to pay tribute to Professor Paul Cloke, who died in 2022. Paul's contribution to rural geography was immense. He founded and edited the *Journal of Rural Studies*, invigorated the subdiscipline with new and significant theoretical perspectives, published widely and influentially on a range of rural topics and, above all, influenced so many through his intellect and kindness. In short, Paul transformed rural geography from a 'Cinderella' subject to one that is now central to geography. Like so many, I benefitted personally from his inspiration, faith, encouragement and support on many occasions. Paul Cloke will be missed, but his influence on rural geography will continue for many years to come.

I would also like to remember Dr Bill Edwards as an inspirational lecturer and caring doctoral supervisor. He encouraged me at all stages of my work and I would not have followed a career in rural geography without his support. I hope this book reflects his passion for rural geography and his ability to communicate to everyone.

Finally, I would like to thank my wife, Ruth, and our children, Elizabeth and William, for their love and support. We've navigated routes across Dartmoor, sampled numerous Devon afternoon teas (cream before jam) and, together, learnt much about the countryside of the South West. In this spirit of applied rural geography, I dedicate this book to them.

Richard Yarwood, Plymouth

Photo Credits

The following photographs have been reprinted with permission from the owners:

Figure 1.1 Second home protests in St Agnes, Cornwall. Photographer: Jori Mundy. 1

Figure 4.1 The view from Leather Tor, Dartmoor. Photographer: Jamie Quinn. 23

Figure 7.1 GWR and the rural idyll. Credit: Great Western Railway. 65

Figure 8.5 A cricket match at Dumbleyung in the Western Australian Wheatbelt. Post for Australian rules football can be seen in the background. Photographer: Les Everett. 82

Figure 13.4 A windfarm near Fordsham, Cheshire, UK. Photographer: Ruth Yarwood. 145

Figure 15.1 A green and pleasant land at the London 2012 Olympic opening ceremony. Credit: Mark Pain/Alamy Stock Photo. 166

Figure 16.1 The Black Farmer produce. Credit: The Black Farmer. 180

Figure 16.2 Khady Gueye with the art installation *Soil Unsoiled* in the Forest of Dean. Credit: PA Images/Alamy Stock Photo. 181

Figure 17.4 Stephen Gough: The Naked Rambler. Credit: Getty/Jeff J Mitchell/Staff. 198

Figure 18.2a Silence at Ramscliffe: MAFF employee monitoring the sedation of the dairy herd, Ramscliffe, Beaford, North Devon, © Chris Chapman 2001. Photographer: Chris Chapman. 205

Figure 18.2b Silence at Ramscliffe: Pithing then spraying the carcasses with dye, Ramscliffe, Beaford, North Devon, © Chris Chapman 2001. Photographer: Chris Chapman. 205

The following photographs are the work of the author Richard Yarwood: Figures 5.1, 5.4, 5.5, 6.3, 7.2, 8.6, 8.8, 13.3, 14.1, 15.2, 15.3, 16.3, 17.3, and 18.1.

Other photographs and figures have been accredited appropriately. Every effort has been made to contact copyright holders for their permission to reprint material in this book. The publishers would be grateful to hear from any copyright holder who is not here acknowledged and will undertake to rectify any errors or omissions in future editions of this book.

CHAPTER

1 Introduction

1.1 Country Lives

The countryside is significant to many but known by few. Most of us live in cities and see the countryside from afar, to be visited occasionally as part of a holiday or trip. It is little wonder, then, that our knowledge of rural places comes from media that frequently portray them as unchanging, tranquil, simple and communal. Rural places are complex and conflicted; they are neither static nor isolated but, rather, places of change and connection. In most countries the countryside accounts for the majority of landuse, is the supplier of food, a place of leisure, a preserve of nature, a valued landscape and, significantly, a home for many. It shapes, and is shaped by, global change, social inequalities, cultural conflict and uneven development. Covid-19, Brexit and the war in Ukraine have demonstrated the significance of the global countryside in the supply of food, fuel and fibre. Now, more than ever, it is important to understand the country and its importance to society and space.

Box 1.1 Country Life

During the Covid-19 pandemic, signs and graffiti appeared in many rural settlements telling second home owners and tourists to 'go home'. At the height of the pandemic, travel for holidays and leisure, including visits to second homes, was illegal in England, but there were concerns that this was being flouted. Elsewhere, in North Wales, concerns were raised that second home owners would swamp limited local services, especially doctors, and bring the virus with them (BBC, 2020). Concern was expressed that some local residents might turn to vigilante methods to discourage or drive away second home owners. Yet tourism is vital to the economies of these areas, and, in Devon and Cornwall, a media campaign #comebacklater encouraged visitors to return at a later, safer date.

Meanwhile, as lockdown eased, estate agents have reported a surge in interest in rural properties, despite protests from some locals that they are depriving them of a place to live (Figure 1.1). Steve Knightley's song 'Country Life' is told from the perspective of a young working-class man in rural England. It highlights grievances

DOI: 10.4324/9780429448966-1

FIGURE 1.1 A protest against second homes in St Agnes, Cornwall. (Source: Jori Mundy)

caused by second-home ownership, class divides, the impact of industrial farming and the closure of services (Yarwood and Charlton, 2009). Its lyrics highlight various social problems faced by people who are socially and economically marginalised in the country.

Lyrics	Hidden Issues
Working in the rain cutting down wood Didn't do my little brother much good Lost two fingers in a chainsaw bite All he does now is drink and fight Sells a bit of grass, hots up cars Talks of travel never gets far Loves his kids, left his wife An everyday story of country life And the red brick cottage where I was born Is the empty shell of a holiday home Most of the year there's no-one there The village is dead and they don't care Now we live on the edge of town Haven't been back since the pub closed down One man's family pays the price For another man's vision of country life My old man is eighty four His generation won the war He left the farm forever when They only kept on one in ten Landed gentry county snobs Where were you when they lost their jobs? No-one marched or subsidised To save a country way of life	Low-paid, dangerous work Probably a long way from a hospital Alcohol abuse, violence, boredom Crime Lack of social and physical mobility Domestic issues Second homes Loss of community spirit Forced local migration to peripheral places with no services Importance of rural idyll Ageing Class divisions Rural politics for elite interests Countryside Alliance Liberty and Livelihood March Foot-and-mouth outbreak of 2001 in the UK

Silent fields empty lanes Drifting smoke distant flames Picture postcard hills on fire Cattle burning in funeral pyres Out to graze they look so sweet We hate the blood we want the meat Buy me a beer I'll take my knife Cut you a slice of country life If you want cheap food well here's the deal Family farms are brought to heel Hammer blows of size and scale Foot and mouth the final nail The coffin of our English dream Lies out on the village green While agri-barons CAP in hand Strip this green and pleasant land Of meadow, woodland, hedgerow, pond What remains gets built upon No trains, jobs No shops, no pubs What went wrong? Country life It's a little bit of country life	Human-animal relations, ethics of farming Consumers distanced from farming Impacts of intensive farming Social/economic loss of small farms Reference to idyllic life that probably never was Environmental impacts of intensive farming and urbanisation, all for profit No services, opportunities or jobs What went wrong? Was it ever right? Nostalgic yearning for something that might never have been, but the song shows the reality

Rural geography offers a way of analysing and understanding these issues. As a subdiscipline of geography, it shares the subject's interests with space and place. Over time, geographers have taken a range of different approaches, or paradigms, to understanding the significance of spatiality to rural society. These range from detailed community studies, through scientific and radical approaches, to recent interests in cultural and post-structural geographies. Rural geography is, therefore, able to reap from a rich crop of thinking to shed light on rural life.

This book introduces this work in four parts: contexts, changes, contests and cultures. Needless to say, there is overlap between these sections and some themes are covered across different chapters. Broad changes in agriculture, for example, are covered in Chapter 5, but the practices and impacts of farming are returned to throughout the book. Likewise, ideas of sustainability, change and inequality run throughout the book.

1. "Contexts" examines the ways in which the countryside has been studied by geographers. The first chapter examines how ideas of community have been closely associated with the country and have influenced how it has been studied. It then moves to discuss how the quantitative revolution attempted to provide tighter definitions of rurality in an effort to make rural studies more scientific. The following chapter examines the notion that we should 'do away' with rurality and, instead, consider the country as no different to the city. The final chapter in this section considers how rurality has been re-considered as a social construction that means different things to different people. Throughout

the section, attention is given to the ways that scale has been applied to rural geography, beginning with the enduring idea of community and ending with more recent global perspectives. The aim of these chapters is not to prescribe particular approaches, nor necessarily advocate the most recent, but, instead, to reveal how different conceptual ideas provide a toolbox for studying different aspects of the country.

2. "Changes" over time and space are addressed in the second section. Drawing on the three elements of Halfacree's (2007) model of rurality, key changes in *rural localities* are discussed in two chapters that examine agricultural change and the post-productive countryside respectively. Attention is then given to *representations of the rural* and the ways in which the countryside has been imagined or socially constructed. Emphasis is given to the way that these representations reflect and affect the geographies of rural change. The final chapter in this section considers the *everyday lives of the rural* by charting some of the social changes in rural places. All the chapters in this section consider the different ways that geographers have studied these issues and illustrate key concepts in rural geography.
3. "Contests" examines some of the conflicts that have arisen from rural change. Sometimes these are overt but, more often, are hidden by cultural expectations and rural politics. Thus, poverty and social exclusion remain prevalent features of rural life but often go unaddressed due to an expectation that people and communities will help themselves. Consequently, the following chapter considers the extent to which policy and governance can empower rural people to address inequality. Housing provides a pertinent illustration of these themes in a chapter that shows how access to a home reflects and maintains social divisions in the country. Indeed, the ability to move in and between places reflects a politics of mobility that is explored in the fourth chapter of this section. The section concludes by considering the importance of sustainability and its significance to all rural issues.
4. "Cultures" draws on the 'cultural turn' in rural geography. The first chapter shows how landscape reflects different cultural viewpoints and how these are reproduced, or 'performed', through daily activities such as walking or farming. The theme of performance is developed in the following chapter, and its significance is discussed in staged, scripted and everyday settings. The next chapter considers Philo's (1992) seminal paper on 'rural others', which argued that rural geography should pay greater attention to diversity in the countryside. Race and disability are used to demonstrate the importance of listening to other voices to reveal the lives of all those living in the country. This theme is developed in the following chapter that examines gender and how women's lives have often been ignored in rural studies and politics. The final chapter examines nature and the significance of non-humans to rural places.

Part I Contexts

Read the following description and try to guess where it is:

> It is 1,572 km^2 and has 300 farms; 8.3 million trees; 1,000 km of footpaths; 850 km of streams, rivers and canals; 13,000 species of wildlife; 2 Special Protection Areas; 3 Special Areas of Conservation; 4 UNESCO World Heritage Sites; 2 National Nature Reserves; 37 Sites of Special Scientific Interest; 142 Local Nature Reserves and 1,400 Sites of Importance for Nature Conservation.

These data were included as part of a bid to recognise London as a National Park City (Greater London National Park City, 2015), something that was achieved in July 2019 (London National Park City, 2022). I neglected to tell you about the 8.3 million people living there, the extent of its built-up area and its role as a global financial centre (FAO, 2014)!

This draws attention to three issues. First, the country and city are not as different as may first be imagined, and indeed, many geographers have stressed that they share many common processes, structures and issues (Hoggart, 1990; Moseley, 1980). Second, it suggests that we associate urban and rural areas with particular characteristics, which are quite arbitrary. Green space and nature are usually associated with the country rather than city, but as London shows, this can be misleading. Finally, rural geographers seem obsessed with the need to define where rural places are. There has been seemingly endless debate on what is or is not rural, as well as a regular flow of papers that seek to (re)conceptualise rurality. By contrast, textbooks on urban geography spend little time defining a city!

Yet it is important to consider the ways geographers have conceptualised rural places because this has significant implications for the way that the country is understood and studied. This book is driven by thinking in human geography. Geography has two main concerns: first, the study of space and place and, second, the relationship between people and the environment. The way that these goals have been approached has varied over time, reflecting different paradigms or

DOI: 10.4324/9780429448966-2

philosophical approaches to the discipline. Rural geography is, therefore, able to reap from a very rich crop of thinking to contextualise and understand rural life.

This section outlines some of the main approaches geographers have used to understand rurality. They are discussed in a broadly chronological order that reflects some of the broader paradigmatic changes that have occurred in geography (Couper, 2014). In doing so, the chapters touch on many topics that are examined in more detail in other parts of the book. The aim of this section is therefore to provide a broad overview of thinking in rural geography.

CHAPTER

2

Where Is the Country?

For many years, the countryside was synonymous with the idea of community, which was often positioned as a more attractive alternative to the anonymity of the city. This chapter traces how ideas of community have been used to inform rural geography and, over time, how these ideas were challenged by efforts to apply more objective, scientific principles to defining and locating rural places.

2.1 Rural Community

The idea of community has been closely associated with rural life (Short, 1992) and, for many years, shaped how the countryside was studied. In the post-war period, social scientists drew on the work of Ferdinand Tonnies (1955, originally published in 1887) to frame their work. Writing in the 19th century, Tonnies distinguished between Gemeinschaft (society), which was based on 'kinship, neighbourhood and friendship', and the more formal Gesellschaft of 'formal contract and exchange'. These ideas were used to identify an urban-rural continuum (Frankenburg, 1966) that recognised the significance of Gemeinschaft in extremely rural areas and Gesellschaft in urban ones (Pahl, 1965b).

Consequently, scholars understood rural places by studying the significance of community to them. These studies were based on months of immersive study and provided rich ethnographic accounts of particular places that revealed the structures, traditions and activities that shaped social relations within them. One influential example of this work was Arensburg and Kimball's (1940) study of 'Family and Community in Ireland' that centred on the town of Ennis (Byrne et al., 2015). Their analysis rested largely on an understanding of the significance of family to community: 'the form of community and lives of its members could not be understood apart from the lives of relations among persons bound by blood, which country people exhibit' (p. xxiv). Thus, a person's gender, age and seniority in a family determined their role and status on a family farm. In times of need, farmers drew on extended family members to assist with farm work. Family was also significant to trade in the town. Young people from farming families were hired by shopkeepers

DOI: 10.4324/9780429448966-3

in the expectation that their families would then use the shop, something that was strengthened through the use of credit to their family members. With the exception of attending church, women were expected to remain at home; by contrast, men's status in the town was conferred at social gatherings in pubs.

These kinds of community studies produced fascinating, holistic snapshots of rural communities at particular times and places (Table 2.2). They also reflected a view that idealised ways of rural life should be preserved. Alwyn Rees' (1950) study of Llanfihangel yng Ngwynfa in Wales not only emphasised the significance of close-knit community but advocated that it should be preserved in light of urbanisation (Day, 1998). He commented that:

> *the failure of the urban world to give its inhabitants status and significance in a functioning society, and their consequent disintegration into a formless mass of rootless nonentities, should make us humble if planning a new life for the countryside.*
>
> *(Rees, 1950: 170)*

This implied that social relations could be 'read from' a particular environment. Thus, urban places were more likely to foster Gemeinschaft relationships, whereas the Gemeinschaft was a product of rural life. Even today, it is not uncommon to hear references to 'urbanites' or 'country people', with an implication that they are inherently different to each other because of the places they live in.

Yet two studies challenged this idea. Herbert Gans' (1962) book 'The Urban Villagers' noted the significance of neighbourliness and community to residents living in inner Boston. At the same time, Ray Pahl's (1965) 'Urbs in Rure' (The City in the Country) examined how rural places were changing as a result of commuting, migration and new forms of employment. He noted how many people who inhabited rural Hertfordshire commuted to work in urban places and spent considerable time there, challenging the binary between city and country life. Pahl (1966: 265) concluded that community studies were an 'uncritical glorifying of old-fashioned rural life' that were descriptive, lacked analytical rigour and without a raison d'etre (Harper, 1989). As a consequence, geographers sought new ways to examine the country.

Box 2.1 Thinking About Communities

Community remains a quandary that has fallen in and out of academic fashion (Liepins, 2000b; Harper, 1989). It has been used to describe three things: a locality, a set of social relations and a sense of communion (Newby, 1986) (Table 2.1).

TABLE 2.1 Three Key Meanings of Community (After Newby, 1986)

Term	Meaning	Example
Locality	The location and description of a settlement	'The Parish of Stillington and Whitton is situated in the Borough of Stockton-on-Tees in North East England, a rural community with a population for the two villages of approximately 1350 people' Stillington and Whitton (2023).

A set of social relations	The interactions that occur between people	'The countryside of England today reflects the wealth of landowners and success of those who can afford to escape the trappings of city life . . . English village communities are small and often eccentric. They are, however, warm and usually welcoming, reflecting local charms of the "English character". The local post office or shop is the communications hub of every village, while the village pub offers an opportunity for new arrivals to get to know local culture and its personalities. But the heart of many rural communities remains a – usually historic – village church around which many community activities – and squabbles – revolve, from village fêtes, parish council meetings to flower arranging' (Orange, 2023).
A sense of communion	A common bond between people	'Building on the legacy of pioneering Scottish and later Dutch immigrants, the community spirit of St. Andrews has inspired active and enthusiastic volunteers for a series of ambitious community initiatives – the building of a fire hall, a community center, a curling rink and a seniors' housing complex – maintaining a thriving rural community into the 21st century. These successes are built on a set of values that puts a premium on self-sufficiency, community spirit, and care for others. By pooling resources, ideas, and talents, the people of St. Andrews have built tangible community services that are unusual for a community of its size (Pop 1,100)'. (Saint Andrews, 2023).

Consider for a moment the communities that you might belong to. There is the community of a university or school, which is usually centred on a particular place or campus. Even within this community, it is possible to identify communities that are based on interests, societies, hobbies, identities, beliefs, accommodation or university courses. When the term ends, you may return to your home village, town or neighbourhood, which brings you into contact with other communities. You may also be part of online communities linked to social media that extend the bounds of community well beyond that of the immediate locality. There is also a sense, perhaps, that a community is more than just a convenient grouping of people, and it helps foster senses of belonging, identity and communality.

For some, these multiple meanings have relegated the idea of community to a 'verbal ragbag' with no analytical value (Cater and Jones, 1989). Community is often used uncritically to describe an ideal, often exclusive, way of life that celebrates an imagined sense of belonging, friendship and support (Valentine, 2001). By describing who is *included*, others are, by definition, *excluded* from communities (Staeheli, 2008). Seen in these terms, the idea of community can be problematic.

Nevertheless, the idea of community remains strong in rural places and is widely used in local decision-making, as a catalyst for local action and to celebrate local life. It does, however, mean different things to different people. It is, therefore, important to appreciate the multiple and contested ways in which community is understood and lived out by different people in the same place (Liepins 2000b) (Chapter 8). Consequently, community remains a problematic but enduring idea that is both a subject of, and approach to, rural geography (Delanty, 2018; Bell, 1994).

TABLE 2.2 The Advantages and Drawbacks of Community Studies

Community Studies	
Advantages	**Problems**
Holistic view of a place	Descriptive
Snapshot of a place in time	Specific
Useful for historical comparisons	Hard to generalise
Can be used to celebrate a way of life	Influenced by idealised views of rural places

2.2 Defining Rurality

One response to the insularity and subjectivity of community studies was to widen the scale of rural geography. Greater consideration was given to the connections between rural places and, in particular, the relationship between the country and the city. Drawing upon the new methods and ideas of the quantitative revolution (Couper, 2014), geographers began to study rural areas using models, maps and data that explained a settlement's function and location in relation to other places (Chisholm, 1962). In some cases, concepts of 'social physics' were applied through 'gravity' models that predicted how the relative size of settlements determined their economic and social 'pull'. Walter Christaller's[1] (1933) Central Place Theory used the size and function of different services to model and predict settlement hierarchies across space (Dacey, 1965). Geographers also drew on ideas from neoclassical economics, such as Von Thünen's (1966) *Der Isolierte State* (The Isolated State), which was originally published in 1826, to model how land rent, or the profitability of particular agricultural activities, influenced land use in the urban hinterland. Although many of the assumptions of these models were simplistic, such as assuming a flat isotropic plain in which people behaved as 'economic men', the quantitative revolution helped to widen the scope of rural geography to the regional level. Rather than studying one community in depth, it considered the relationship of communities with each other.

There were also many attempts to define rural places using statistical analysis of secondary data. One of the most influential was Paul Cloke's (1977) Rurality Index that used a range of census data[2] to distinguish between different types of rural places in England and Wales and class them as 'extreme rural', 'intermediate rural', 'intermediate non-rural' and 'rural'. Others mapped rurality using different data sets or approaches. These included measures of agricultural function, position in the urban hierarchy, regional importance or the extent of built development (Halfacree, 1995).

Measures of population have also been used to define the difference between urban and rural areas, although these vary widely between different countries. In Sweden, a settlement with over 200 people is considered urban (Statistics Sweden, 2019), whereas the number is 10,000 in the UK (Office for National Statistics, 2016). These population measures are therefore usually combined with other, contextual indicators. For example, the European Union uses a three-step approach to define an urban-rural typology that combines population density, a regional classification and the presence of cities (Eurosat, 2020b) (Box 2.1). The 'official' definition of rural areas in England uses measures of population to identify four settlement types: urban (population over 10,000), town and fringe, village, hamlet and isolated dwellings (Office for National Statistics, 2016). The wider areas in which these settlements are located are then defined as 'sparse' or 'less sparse', depending on whether 80% or 50% of their population live in rural places (Figure 2.1). The US Census Bureau refers to rural areas as 'any population, housing, or territory NOT in an urban area'. Urban areas refer to 'urbanised areas', which have a population of over 50,000, and 'urban clusters', which have populations between 2,500 and 50,000 people (United States Census Bureau, 2020).

Box 2.2 The European Union's Definition of Rural Areas

There is a three-stage approach used by the EU to define rural places (Figure 2.2).

1. Defining Rural

The first step is to identify populations in rural areas: rural areas are all areas outside urban clusters. Urban clusters are clusters of contiguous grid cells of 1 km^2 with a density of at least 300 inhabitants per km^2 and a minimum population of 5,000.

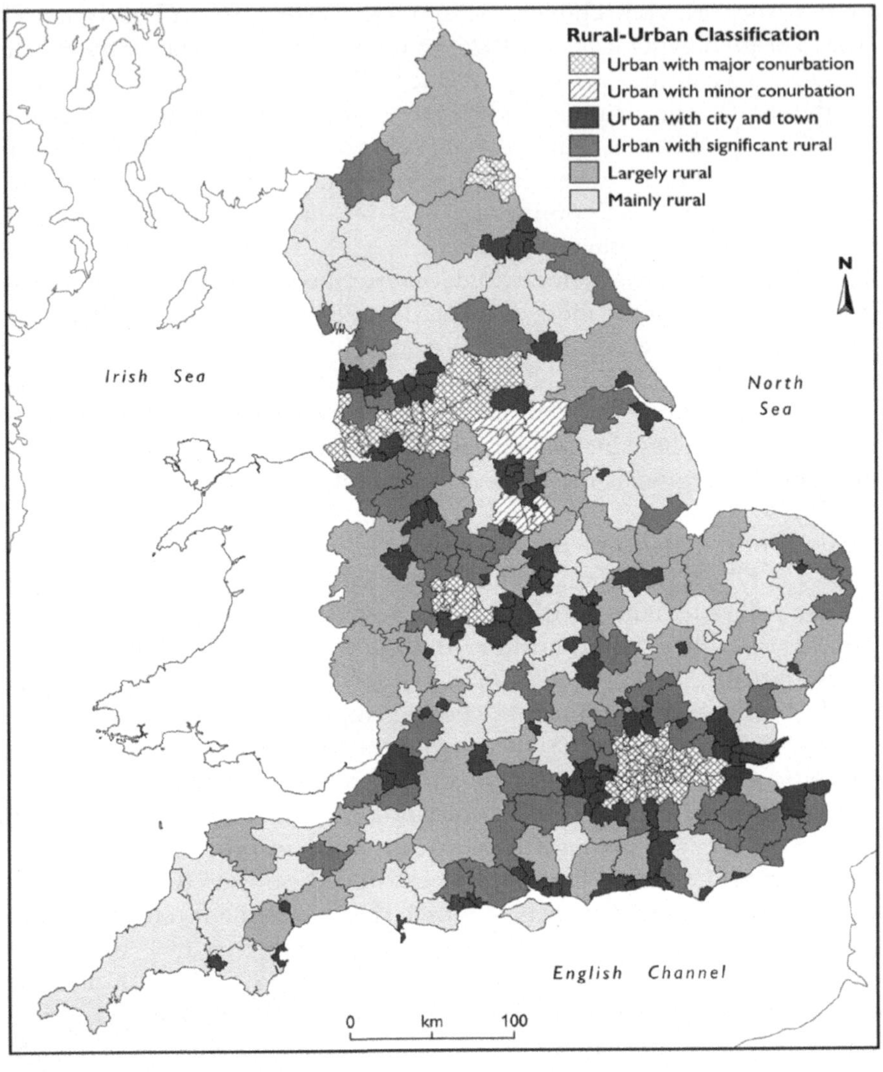

FIGURE 2.1 Rural areas in England.
Source: Office for National Statistics (2016)

2. Classification of the Regions

In the second step, regions are classified according to the share of population living in rural areas. They are:

- 'Predominantly rural' if the share of the population living in rural areas is higher than 50%;
- 'Intermediate' if the share of the population living in rural areas is between 20% and 50%;
- 'Predominantly urban' if the share of the population living in rural areas is below 20%.

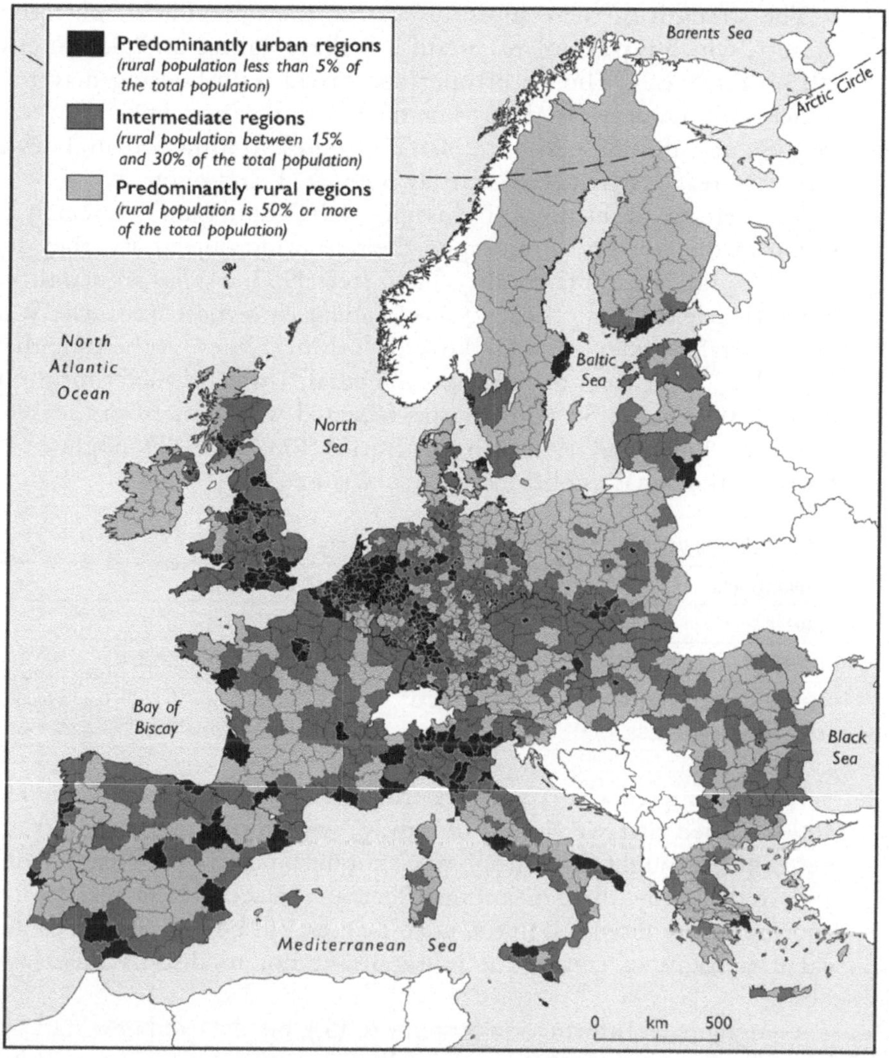

FIGURE 2.2 Rural and urban areas of the EU (Eurostat, 2020b) (n.b. the UK is no longer a member state).

Regions smaller than 500 km^2 are combined with one or more of their neighbours.

3. Presence of City

A predominantly rural region which contains an urban centre of more than 200,000 inhabitants that accounts for at least 25% of the regional population is defined as 'intermediate'. In turn, an intermediate region that contains an urban centre of more than 500,000 inhabitants and makes up at least 25% of the regional population is classed as 'predominantly urban'.

Source: Eurosat (2020b)

The strength of these measures is that they provide an objective measure of rurality, which is particularly useful in policy-making and planning (Gallent et al., 2015) (Table 2.2). Thus, a planner concerned with housing may refer to specific settlement boundaries to determine the location of new housing; a transport officer may use measures of sparsity to plan bus routes; funding for rural development may focus on areas designated as 'sparsely populated' and so on.

Yet there are problems with this approach. There is no one widely accepted definition of the country, and despite their seeming objectivity, they reflect ideas of what 'should' constitute rurality. Halfacree (1993: 24) has argued this involves 'trying to fit a definition to *what we already intuitively consider to be rural'*. When reflecting on his rurality index, Paul Cloke said, 'I think I knew at the time that by selecting a number of variables to represent the rural, I was pre-determining the outcome' (Cloke, 1994: 156). So while it was repeated with data from the 1981 (Cloke and Edwards, 1986) and 1991 census (Harrington and O'Donoghue, 1998), Cloke's Rurality Index has not been reproduced since.

TABLE 2.3 The Value of Statistical Definitions of Rurality

Advantages	Drawbacks
Objective	Subjective
Useful for decision-making	No two are the same
Helpful for measuring, mapping, monitoring and comparing rural places	Descriptive rather than analytical

This prompted two responses from geographers. First, Keith Hoggart (1988, 1990) argued that we should 'do away' with measures of rurality. Second, other geographers sought to examine what rurality means to different people and groups and, in turn, how these meanings affect rural places (Halfacree 1993). Both of these approaches are discussed in the next chapter, but before moving away from statistical definitions of rurality, it is important not to dismiss them as unhelpful or obsolete.

Geographical Information Systems (GISs), big data and new statistical techniques have allowed rural places to be modelled with ever-growing sophistication. Statistical approaches have been used to map and analyse, amongst other things, suicide rates (Cheung et al., 2012; Dorling and Gunnell, 2003), access to health care (Arcury et al., 2005; Lovett et al., 2002), incidence of crime (Ceccato and Dolmen, 2011) and deprivation (Higgs and White, 1997; McEntee and Agyeman, 2010) in rural places. Quantitative approaches continue to be particularly important in policy-making, with spatial analysis being utilised to identify regions, localities and pockets of the countryside that warrant policy interventions (Gallent et al., 2015; Ward and McNicholas, 1998b). For these reasons, spatial analysis will continue to play an important role in rural geography, with advances in technology and mathematics lending power to these approaches.

Geographers continue to draw upon various 'urban' and 'rural' statistics to contextualise their work or make a case to support an argument. Indeed, without a basic knowledge of what rural areas are 'like', it would be difficult to study their geographies. As such, this book continues to refer to data that describe urban and rural places, which, in turn, are based on various measures of rurality. Statistical measures of rurality, despite their shortcomings, provide important platforms for more theoretically informed research. The crucial thing is to be aware of how they are defined and the influence this has on the ways that data are presented.

Notes

1 Other geographers were to later question Christaller's role in the Nazi Party and how he used central place theory to shape occupied territories and their people. (Barnes, 2015).

2 Occupancy rates; commuting out; female population, aged 15–44; household amenities; population density; occupational structure; population over 65; and distance from urban areas of 500,000 people. Harrington V and O'Donoghue D. (1998).

CHAPTER

3

Doing Away with Rurality?

One of the first articles published in the *Journal of Rural Studies* had the rather surprising title of 'Let's do away with rural' (Hoggart, 1990). Happily for the journal, which continues to publish world-class research on rural issues, Hoggart was not calling for an end to rural geography. Rather, he wanted to end the quest to define rural places and, instead, place a stronger focus on the processes that shaped them (Hoggart, 1988).

In urban studies, the orthodoxy of quantitative geography had already been challenged by David Harvey's (1973) book *Social Justice in the City*, which argued geographers should be concerned with analysing and challenging social inequalities. Harvey drew upon Marxist theory to understand how capitalism shaped urban places and the ways in which its uneven development relied upon and maintained unequal social relations.

Understood in this way, rural society, like its urban counterpart, was simply the outcome of a set of industrial relations and, in many rural places, this industry was agriculture (Newby, 1979a).

Box 3.1 The Political Economy and Rural Studies: A Basic Introduction

David Harvey's (1973) seminal book, *Social Justice and the City*, introduced Marxist thought to geography and used it to examine the relationship between capitalism and space. Harvey's work not only sought to describe and explain change but argued geographers should 'mobilise the powers of thought to formulate concepts and categories, theories and arguments, which we can apply to the task of bringing about a humanizing social change' (p. 145).

Capitalism is incredibly sophisticated and its analysis has led to a wide and complex body of work. Simplifying this literature, it is possible to

DOI: 10.4324/9780429448966-4

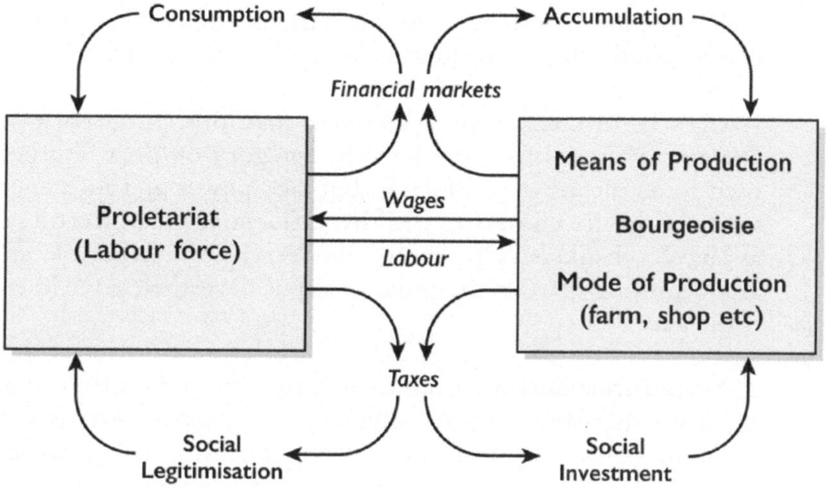

FIGURE 3.1 A summary of capitalism.
Source: A simplification of Harvey (1989)

summarise that capitalism relies on three things: the accumulation of capital, unequal class relations and state intervention (Figure 3.1).

1. Accumulation for accumulation's sake. The purpose of capitalism is to accumulate further capital. This can refer to monetary profits, but capital can also be accumulated in many other ways. For example, money can be invested in fixed assets, such as property, that can go up in value.

The way that capital is accumulated is referred to as the *mode of production*. Thus geographers have studied the significance of Fordist modes of production in the countryside. Fordism refers to mass production and was crucial in driving productivist forms of farming in the post-war period (Chapter 5). Since the mid-'80s, post-Fordism has emerged as a new mode of production that is based on forms of flexible accumulation and state deregulation (Cloke and Goodwin, 1992)

The *means of production* refers to the ownership of assets, such as a farm, land, business or even a house, which can accumulate capital. Not everyone has access to these, which leads to class conflict.

2. Class relations. Capitalism relies on two basic classes of people: those who own the means of production (the bourgeoisie) and those who must sell their labour for wages (proletariat). Yet given the need to accumulate capital (see item 1), the bourgeoisie will always pay less than the value of the good they are producing. Thus, if a crop sells for £10,000, the labourer will be paid much less than this in order to make a profit.

Nevertheless, wages allow people to consume (buy) goods and thus allow the accumulation of capital by the bourgeoisie. The latter may seek to reduce labour costs (to increase profit) by moving production to places where wages are low or by employing migrant labour at lower costs.

Class is, of course, more complex than this. Property ownership or self-employment lifts some into the category of 'petit bourgeoisie', who own some means of production but lack power and influence. By contrast, the service classes are wealthy, influential and powerful people, such as lawyers, bankers or teachers, who 'service the needs of capital'. They are technically part of the proletariat but nevertheless wield considerable influence.

3. State intervention. Both the bourgeoisie and proletariat pay taxes to the state that, in turn, are invested in forms of infrastructure to aid accumulation (roads, rail links, etc.) or forms of *social legitimisation*. This refers to the provision of education, housing or welfare to ensure a healthy, well-trained workforce to meet the needs of capital. As such, governments and their policies are part of the capitalist system rather than independent of it.

To better understand the relationship between capitalism and the country, many geographers adopted a political economy approach. Driven by Harvey's (1973) book, political-economic theory was used to analyse how capitalism shapes social, economic and political relationships (Box 1). Whereas community studies examined particular villages in great detail, geographers became more concerned with how they were affected and shaped by wider changes in the structure of society. As Newby (1986: 212) commented, 'although many studies were pursued with a local setting, they were not concerned, first and foremost, with documenting the distinctiveness of particular localities or rural communities'. Consequently, the identification and definition of rural areas became of secondary importance as 'causal processes do not stop at one side of an urban-rural divide'! (Hoggart, 1988: 36). Rurality *per se* was not as significant as the political, social and economic structures that shaped it.

Instead of reading social relations from an arbitrary urban-rural continuum, rural society was explained by understanding the uneven impact of global capitalism on different places (Figure 3.2). Attention was given to identifying key regimes (Box 3.2) and the ways that transitions between them took 'different forms and proceeded at different scales at different times in different rural areas' (Cloke and Goodwin, 1992: 327). This approach produced holistic accounts of change that linked global changes in production and consumption to social changes in particular places (Chapters 5 and 6).

These political-economy approaches help to identify and explain inequalities in rural society. For example, rural poverty could be explained by the ways capitalism exploits people for cheap labour in peripheral areas in order to make profit elsewhere (Tickamyer, 2006) (Chapter 9). Thus, the wealth of the Global North largely depends

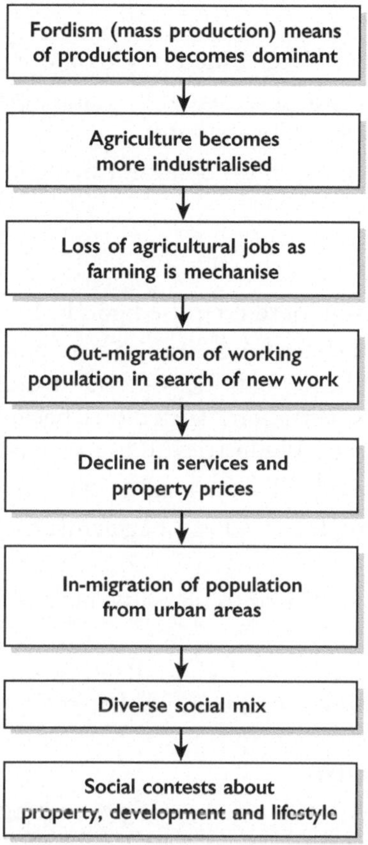

FIGURE 3.2 A simplified account of rural structural change.

on an unequal set of trading relationships with the Global South (Rignall and Atia, 2017; Wilson and Rigg, 2003; Tickamyer, 2006). Some gender inequalities reflect a historical division of labour that required women to do unpaid work in the home and on the farm (Whatmore, 1988) (Chapter 17). There is a continued shortage of housing because building more would lower existing property prices (Shucksmith, 1990) (Chapter 11). In short, the shift away from quantitative to radical geography led to a more critical and deeper understanding of the country.

Box 3.2 Changing Regimes of Accumulation

A regime of accumulation refers to a stable set of social, economic and political relationships that characterise how capitalism operates (Hall and Barrett, 2018). Although there is debate about the extent to which regimes change, geographers have referred to three regimes characterised by different modes of production and consumption in the countryside (Woods, 2004: 1034).

1. Paternalism (pre-1930s):

- ownership of land and resources determined wealth and status;
- employment in rural localities was dominated by single, primary industries, often structured around local estates;
- local elites provided welfare and accommodation through paternalistic forms of charity.

2. Fordism (1930s–1980s):

Fordism refers to the mass production of standardised goods. In rural areas:

- Fordist principles were applied to farming through state policies that encouraged the production of large quantities of food through intensification, specialisation and concentration. Agriculture became more closely connected to agri-industrial industries and supermarkets;
- welfare and housing were provided by the local governments;
- the industrialisation of farming led to a loss of agricultural jobs and out-migration;
- manufacturing plants were established to take advantage of cheap, rural workforces;
- increased migration to the countryside by middle-class migrants with the power to consume and 'buy into' rural lifestyles.

3. Post-Fordism (1980s onwards)

Post-Fordism is characterised by flexible forms of production and consumption. In the countryside;

- consumption rather than production became the dominant economic activity;
- there was growth in the service sector and amenity countryside;
- the state withdrew from some services, to be replaced by voluntary and private sectors;
- new forms of governance drew together public, private and voluntary sectors into partnerships;
- the countryside is now connected to distant places through global flows of production and consumption.

TABLE 3.1 Advantages of Political Economy Approaches

Advantages	**Drawbacks**
Holistic analysis of rural issues	A 'big picture' can hide local differences
Develops a critical understanding of rural areas; identifies ways of prompting change	Concerned with theory over policy
Emphasises analysis over description	Sometimes lacks an empirical dimension
Looks beyond rurality	Fails to appreciate what rurality means to different groups
Radical, critical thinking	Ideological

Capitalism has also forged connections between rural places across the world to the extent that many geographers now refer to the 'global countryside' (Woods (2007). The term draws attention to the increasingly significant connections between rural places across the world (Table 3.2). Rural places are no longer, if they ever were, discrete, bounded and identifiable but, rather, are:

> *knitted-together intersections of networks and flows that are never wholly fixed or contained at the local scale, and whose constant shape-shifting eludes a singular representation of place.*
> *(Woods, 2007: 499)*

TABLE 3.2 Facets of the Global Countryside

Feature	Characteristics
1. Extended commodity networks	Extended food chains, characterised by longer 'food miles'; global trade is supported by deregulation and tariff reform (Chapter 5).
2. Corporate concentration	Transnational corporations dominate key sections of the rural economy and are aligned with globalised chains of food production (Chapter 5).
3. Supplier and employer of migrant labour	Food production in the Global North relies on seasonal migrant labour, often from rural areas of poorer countries (Chapter 12).
4. Flows of tourists: national and international	Some tourist sites have gained a global reputation, leading to increased visitor numbers from across the world (Chapters 6 and 14).
5. Non-national property investment	International investment occurs via land grabs or buying properties for holidays or investment (Chapter 6).
6. Commodification of nature	Nature, animals and the environment have been reimagined and exploited through global investments in tourism, resource exploitation and other economic developments (Chapters 13 and 14).
7. Impact on the landscape	Globalisation has led to large-scale and significant changes to some landscapes, evidenced by deforestation or environmental degradation, as well as the introduction of new crops and animals to the detriment of local species (Chapter 13).
8. Inequality and social polarisation	While globalisation has benefitted some groups, others have lost land and employment or are priced out of rural housing markets (Chapters 9 and 11).
9. New sites of authority	National parks, economic regulation and international agricultural policies are imposed on rural places that disempower local people (Chapter 10).
10. Contest	These transformations have led to resistance and a new politics of the rural, in which different ideas of rurality are contested (Part III).

Source: Woods (2007)

Globalisation is not a one-way process. Although political economy approaches aimed to paint a 'big picture' of change, change is not a one-way process and local geographies, histories and circumstances mediate the impact of change

(Massey, 1991). Cloke and Goodwin (1992) used the term 'localised structured coherences' to describe how places are formed and shaped by a combination of exogenous changes and endogenous social and cultural circumstances.

Rurality was therefore never quite 'done away' with. Instead, geographers recognised that the distinctiveness of rural places matters and that local social relations affect the ways in which new rounds of capital impact the countryside (Halfacree, 1993; Harvey, 1989). Cloke and Goodwin (1992) referred to a 'cultural glue' that held places together and shaped the process of change. As a consequence, geographers took a renewed interest in the way the countryside was imagined or 'culturally constructed' by different people in different places. The significance of these ideas is explored in the following chapter.

CHAPTER

4

Geographies of the Rural

FIGURE 4.1 The view from Leather Tor, Dartmoor.

Figure 4.1 was taken on Dartmoor, a national park in the South West of England. In the foreground is Leather Tor, one of 365 tors or granite outcrops in the park (Ringwood, 2013), and in the background is Burrator Reservoir, which supplies drinking water to the nearby city of Plymouth. Surrounding the water are forestry plantations and farmland. You may think that the landscape looks wild, a description often applied to Dartmoor, but the artificial reservoir and evergreen trees point to the dominant influence of people on the landscape.

I took the picture while out walking to capture a picturesque view. Picturesque, incidentally, describes a style of landscape painting that has a foreground (the tor), middle-ground (the reservoir) and background (distant hills) and has strongly influenced how we value a landscape. But how might the picture have been read by other people? A farmer, for example, might consider the suitability of the land for grazing sheep; a water board official, the levels of water; a Forestry Commission officer, the condition of the trees; a soldier, how to cross the ground unobserved;

DOI: 10.4324/9780429448966-5

and so on. But by taking a picture of the view, I was reflecting and contributing to the idea that a national park is a place of leisure and wild beauty. As these examples show, the country means different things to different people, or as Mormont (1990) puts it, rural is a category of thought.

4.1 Rurality as a Social Construction

Keith Halfacree (1993) suggested that rurality was best understood as something that was 'socially represented' or 'culturally constructed'. Central to his argument was the idea that '"the rural" should not be seen as a singular object of discourse' (p. 32). In other words, the term 'rural' has come to signify different meanings and ideas; it is subjective rather than an objective. Halfacree advocated understanding rurality as a discourse or set of communications that reflect and reinforce particular sets of ideas. Rather than having a fixed meaning, rurality emerges through the ways it is imagined, communicated and lived out on a daily basis (Chapter 7). These are often perpetuated to assert and maintain power relations. As John Short (1991) argued, our task is not to determine which constructions are 'true' but, rather, to examine whose truth it is and why this matters.

TABLE 4.1 Eight Dimensions of Rurality

Characteristic%		Both/neither		Characteristic%
Relaxation	88	12	0	Stress
Tradition	81	19	1	Modernity
Healthiness	81	1	1	Unhealthiness
Safety	65	28	7	Danger
Natural	64	25	12	Man-made
Community	59	34	7	Loneliness
Simplicity	47	46	7	Sophistication
High Status	41	40	19	Low Status

Source: Halfacree (1995)

Thus, an idealised vision of country living has prompted people to move to the countryside (Chapter 8); rural heritage may be used to protest against built development (Chapters 10 and 14); ideas of community are used to brush over inequality; and idyllic visions of the countryside have continued to hide social problems (Bunce, 1994; Yarwood, 2005a; Vepsäläinen and Pitkänen, 2010; Cloke et al., 1995a; McLaughlin, 1986a; Little and Austin, 1996; Shucksmith, 2018; Matthews et al., 2000) (Part III). For example, Halfacree (1995) interviewed 113 migrants to parishes in Lancashire and Devon, England, and revealed that rurality was commonly associated with many idyllic characteristics (Table 4.1). While Halfacree cautions that residents are not 'naïve advocates of a mythical rural world' (p. 19), he nevertheless notes that these idyllic views contributed to migration.

While the rural idyll is a powerful and significant discourse, it is not the only way that the countryside is viewed. The countryside is sometimes portrayed as backward, hostile, underdeveloped and isolated: an anti-idyll (Short, 1991). More problematically, those that do not 'fit' into an idyllic country vision, such as the poor or travellers, can be physically, socially or culturally marginalised in the countryside. In a seminal paper, Chris Philo (1992) challenged geographers to listen to marginal or 'other' discourses that were rarely heard in the country. A range of studies followed that sought to re-examine the countryside from the perspective of 'other groups' (Cloke, 2003; Cloke and Little, 1997) and draw attention to people, practices and places that had often been neglected from rural studies (Chapter 16).

Box 4.1 Post-Rural Geographies

The term 'post-rural' recognises that rurality is reflexive (Murdoch and Pratt, 1993: 425). In other words, different groups of people recognise that rural places are imagined but, nevertheless, continue to behave in ways that reflect and maintain these ideas. A tourist, for example, may realise that a visit to a rural tourist attraction presents a 'packaged' and partial view of the countryside, but they are content to accept this in return for a day out (Urry and Larsen, 2011). Similarly, a move to the country may be prompted by an idyllic image of the rural life, even though the migrants may recognise that this may not match the realities of rural living.

Post-rural geography reflects a wider post-structural turn in geography. This paradigm refutes the idea that society is made up of permanent 'structures' or 'identities' but, rather, is continually being formed and reformed in different ways. It challenges how habitual categories – such as class, gender or rurality – are not 'fixed' but, rather, are discursively created through everyday activities and practices. These are not neutral but, instead, are constructed by different historical, political and spatial conditions. Post-structural/post-rural approaches have led to consideration of the ways that rurality is socially constructed and contested, how these ideas are configured through daily activities or 'performances' and the significance of relational networks to the creation of rural space.

Performance

Dominant constructions of the countryside reflect and maintain structures of power and influence, but they are maintained or challenged through daily actions and everyday lives (Woods, 2010). The idea of 'performance' describes how particular ideas are re-enacted or reproduced through particular actions (Chapter 15). For example, the skill of ploughing or the daily routines of keeping livestock maintain what is understood as 'good' farming practice (Burton, 2004). Similarly, the act of going for a walk in the countryside confirms that the countryside is a place of leisure and reflects culturally informed ways of seeing and behaving in rural places

(Edensor, 2000; Wylie, 2005). Alternatively, the way that soldiers train to see, live and fight reflects that the countryside can be a challenging wilderness (Woodward, 2000). These investigations have led geographers to examine the ways the countryside is experienced through senses, bodies and emotions. This moves beyond how the country is *represented* and, instead, reveals how it is *experienced* in the here and now (Thrift, 2008).

TABLE 4.2 Post-Rural Geographies

Advantages	Drawbacks
Sensitive to local cultures and difference	Depoliticised and does not focus on the structures that create a difference
Shows the importance of images, signs, emotions and the non-representational	Ignores the empirical 'realities' of the countryside
Challenges binaries and conventional meanings of rurality	Lacks a 'big picture' view of change

4.2 Hybrid Ruralities

Rural areas do not just reflect human activity. People, animals (wild and domestic), birds, buildings, trees, plants, water and many other forms of 'nature' all combine to create rural places (Cloke and Jones, 2001) (Chapter 18). Different technologies, people, institutions, knowledge, performances and practices come together in particular ways in particular spaces to create specific rural places (Murdoch, 2006; Whatmore, 2002).

Consider, for example, how a farm is composed of animals (such as cattle), technologies (milking parlours), knowledge (acquired over centuries and/or from formal training), regular activities (milking), and economic networks that stretch well beyond the farm gate. In this way, the farm might be thought of as an 'assemblage' of different things and actions (Holloway et al., 2014). Rather than focusing on the significance of a particular agency, it is important to recognise its importance *in relation* to other agencies. How, for example, do cattle behave in relation to an automated milking parlour?

It is therefore possible to consider the countryside an outcome of a series of networks that are drawn together in particular places (Heley and Jones, 2012; Murdoch, 2006). This is useful for three reasons.

First, it suggests that the countryside is 'more than human' and is 'co-constructed by humans and non-humans, bound together in complex inter-relationships' (Murdoch, 2003: 264). This approach has been applied to critique binaries of 'nature' and 'society' and provide clearer understanding of the ways in which humans and non-humans are entwined with each other (Holloway et al., 2014, Sellick and Yarwood, 2013).

Second, hybridity has also challenged the idea that scale reflects a hierarchical progression from the local to international (Wilson, 2012). Instead, processes can operate at different scales at the same time (Marston et al., 2005; Herod, 2011). A farmer, for example, may work in the private space of the home (farmhouse), engage with the local community, follow policies made at the national level and sell food as part of a global commodity chain (Kelly and Yarwood, 2018). The term 'transrural' (Askins, 2009) has been used to recognise how actions in local places can connect to other places across the world (Chapter 16).

Finally, networks emphasise how different and distant places are connected through flows of people, ideas, technology and media. Murdoch argues (2003: 274) that 'a focus on networks and fluid spaces will disrupt the notions of easily demarcated and fixed rural spaces'. Seen in this way, rural places are not bounded and isolated but, rather, are connected and fluid. Rural places are never 'fixed' but, instead, are constantly being changed and shaped by networks that extend well beyond the immediate locality. This is leading to an increasingly connected and globalised countryside. Indeed, Woods (2007) considers that globalisation is an outcome of three interconnected forms of hybridisation:

1. Hybrid and multi-stranded globalisation: globalisation is not universal in its nature or impact. Various elements of globalisation (Table 3.2) reflect the outcome of different hybrid networks that combine uniquely in different places.
2. Globalisation as hybridisation: globalisation is itself a hybrid process that combines global change and local circumstances. Its outcomes are not inevitable, but rather, 'local actors engage with global networks and global forces to produce hybrid outcomes that are fundamental to the reconstitution of place in the globalizing countryside' (Woods, 2007: 497). Globalisation is best thought of as an incomplete series of events rather than a finished product.
3. More-than-human globalisation: globalisation mixes human and non-human elements. Nature and animals are incorporated into global, for example, through global tourism (Cloke and Perkins, 2005) or international exports of farm animals (Tonts et al., 2010). As the Covid-19 pandemic illustrated vividly, this also extends to viruses.

Work on relational networks has also drawn attention to other scales of analysis. Some geographers have concerned themselves with microscopic scales, examining how diseases and viruses cross political borders as well as the boundaries between species to impact on rural places (Hinchliffe et al., 2013; Enticott, 2008a). Such work is pertinent given the impact of the Covid-19 pandemic. More recently, the scales of the body and the home have led to new accounts about the ways that people engage physically and emotionally with rural spaces (Chapter 15) (Little and Leyshon, 2003).

4.3 Conclusions: The Threefold Model of Rurality

The first four chapters of this book have sought to outline some of the approaches taken by geographers to conceptualise what is meant by a rural area. The aim is not to privilege one paradigm over another but, rather, to highlight how different approaches provide different ways of understanding rural places. Rather like ingredients in a kitchen, particular ideas of rurality can be used to make different dishes. And, as in cooking, particular ingredients may be more appropriate than others for making particular dishes: just as pepper would be unsuitable in a trifle but tasty in a stew, different ideas of rurality should be used appropriately. Thus, functional measures of rurality would be helpful to map or chart rural change but would not help to explain why it occurred or how it is experienced.

There are different approaches to understanding rurality, and each has its own advantages and drawbacks. It is unlikely that a universally accepted definition of rurality will ever be achieved and, in any case, is not desirable. Indeed, the different ways of conceptualising rurality, and the debates surrounding them, point to the rude health of rural geography and the richness of its different intellectual traditions. This book uses a flexible rather than dogmatic approach to conceptualising rurality, using different ideas of rurality to explain different aspects of countryside change.

Nevertheless, there is some benefit in bringing these different approaches together. Keith Halfacree's (2007) threefold model of rurality is a major contribution to thinking about rurality (Figure 4.2). Drawing on ideas by the French sociologist Henri Lefebvre (1974), Halfacree argues that rural places can be understood in three ways:

1. Rural localities – how distinctive aspects of rural space are the outcome of economic and social restructuring and, in particular, new forms of production (growing or making products) and consumption (buying products).
2. Formal representations of the rural – how rural interests are represented by key stakeholders, such as planners, politicians and other decision-makers. Here it is instructive to consider how rural areas are socially constructed or represented by different groups. While dominant ideas are important to understand, there is also scope to examine alternative or contested views of rurality.
3. Everyday lives of the rural – the diverse and subjective ways in which people live their lives in rural places. These may be viewed as the outcome of restructuring, but different performances of rurality can enforce or challenge dominant processes of change or representations of rurality.

The relationship between these elements produces rural places that are 'unified', 'contradictory' or 'chaotic' (Halfacree, 2007):

1. Congruent and unified. 'All elements of rural space cohere in a relatively smooth, consistent manner' (p. 127). Examples of this can be found in localities where productivism is dominant, such as the Western Australia Wheatbelt (Jones and Tonts, 1995),

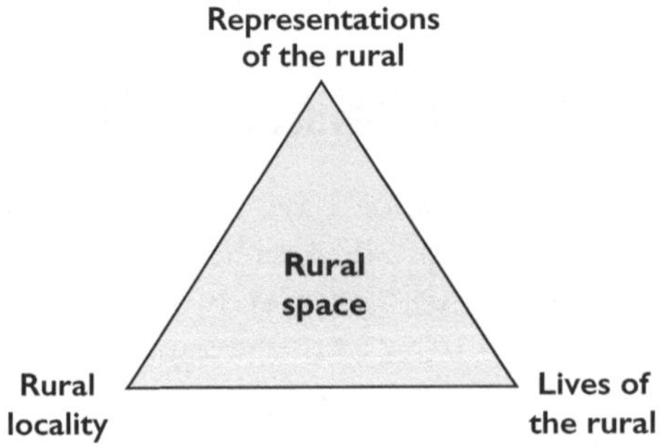

FIGURE 4.2 Halfacree's threefold model of rurality.
Source: Halfacree (2007)

East Anglia in the UK or the American prairies. Here localities have been changed economically and environmentally by intensive farming practices, which have in turn been supported by legislation and policy that support a view that the countryside is primarily a place for the production of food. Everyday life is structured around farmers and farming through employment, social and policy networks. Equally, the post-productive countryside might also be congruent and unified. Take, for example, locations such as Queenstown in New Zealand, Cheddar Gorge in England or Yosemite in the USA. These places and others like them have been transformed by tourism to the exclusion of other activities. They have been reimagined through advertising campaigns that promote the tourist gaze and exclude other perspectives. Planning and infrastructure are almost entirely geared towards managing tourists' experiences. Employment is largely in the service sector; others, such as farmers, play a relatively minor role in social and political structures.

2. Contradictory and disjointed. 'There is tension and contradiction within/between elements of rural space but an overall coherence holds' (p. 128). Rural space is disjointed in places where dominant activities are being challenged by new ones. In South Devon, for example, mixed family farming is significant but so, too, are tourist activities and in-migration from middle-class residents. The daily lives of people are more diverse: some are dependent on agriculture and others on jobs outside the locality. Contradictions derive from different ways in which rurality is imagined and the relative power of lobby groups to enforce these ideas. These are revealed in planning disputes over 'new housing, roads, windfarms, quarries, extraction pits, waste dumps and other similar land uses that are the antithesis of some conceptions of rurality, but which can be accommodated within others' (Woods, 2006, 580).
3. Chaotic and incoherent. 'There are fundamental contradictions within/between the elements of rural space . . . fundamentally conflicting ruralities co-exist. Difference is again readily apparent' (Halfacree, 2007: 128). These places may reflect overt conflicts over the nature of economic change, rural ideologies or the interests of power. They might include places influenced by back-to-the-land movements, self-sufficiency, minority interests, indigenous groups or alternative beliefs (Tonts, 2001) that conflict with wider, hegemonic society. For example, disputes may occur over trespass, ownership, planning regulations or whether activities, such as raves, are compatible with rural life. These may be resolved through the planning system, legal challenges and, occasionally, violent conflict or everyday hostility.

The significance of Halfacree's threefold model is that it draws on structural, social constructionist and performative ideas to understand and conceptualise rural space. As such, it provides a flexible 'middle' theory to link 'the concerns of the cultural turn with those of political and economic mainstream' (Cloke, 2006a: 26). Consequently, it has become widely adopted in rural studies (Woods, 2011b) and is used in the next section of this book to consider how and why the countryside is changing.

Part II Changes

The previous section traced some of the ways in which geographers have understood rural change. This part builds on the previous chapter by utilising Halfacree's (2007) threefold model of rural space to trace significant changes in the countryside. The first two chapters – on agricultural and economic change – trace how rural localities have changed; the next chapter considers how rural places are represented, particularly in idyllic ways; and the final section considers social change. All four chapters can be read in a stand-alone way but, taken together, draw together Halfacree's (2007) three aspects of rurality.

DOI: 10.4324/9780429448966-6

CHAPTER

5

Farming and Food

5.1 Introduction

Many of us reading this book will be well-fed and have access to a plentiful supply of food, at least if we live in the West. Yet the Covid-19 pandemic led to empty supermarket shelves, shortages of staple products, threats to supply chains, a lack of seasonal workers to pick and process food, children going hungry as schools closed, wider use of foodbanks and the mobilization volunteers to distribute food to vulnerable people and communities. If this came as something as a shock, it highlighted that food is taken for granted by some but a tenuous commodity for many.

This precarity is further exacerbated by a series of global conditions that are challenging our ability to produce food in a plentiful, safe and sustainable way (Evans, 2013) (Table 5.1). The FAO (Food and Agriculture Organization of the United Nations (FAO), 2009) has estimated that food production will need to increase by 70% by 2050 in order to feed a predicted world population of 9.1 billion people. This requires an expansion of cereal harvests from 2.1 to 3 billion tonnes per year and for annual meat production to increase by 200 million tonnes. In part, the latter is being driven by a growing middle class in the Global South who have a desire for improved, meat-based diets. In 2013, 70 billion animals were killed for food, which is a ninefold increase in 50 years (Weis, 2013).

Against this backdrop, fewer people are working in farming, and there has been an increase land used in non-food production, including growing crops for biofuels or dedicating land for renewable energy. Further, agriculture accounts for around a quarter of global carbon emissions, contributing to climate change that, in turn, threatens the sustainability of the global food supply (Poore and Nemecek, 2018).

DOI: 10.4324/9780429448966-7

TABLE 5.1 Global Challenges to Agriculture (Evans, 2013)

Challenge	Reasons for Agricultural Concern
Population growth	Levels of food production need to increase to feed a growing population.
Nutrition transition	Growing middle classes in some countries are prompting growth in meat consumption.
Urban growth	Urbanisation is putting pressure on the land available for agriculture. In 2009, the number of people living in cities surpassed those living in rural areas for the first time. By 2050, the urban population is expected to be 68%.
Labour/skills shortage	Relatively fewer people work on the land, reducing agricultural skills, knowledge and labour.
Climate change	The climate emergency will lead to warmer temperatures, an increase in extreme weather events and, particularly in the Global South, less favourable growing conditions.
Water shortages	Agriculture in many places relies on unreliable or depleted water resources that will become further threatened by climate change
Energy demands	Agriculture is heavily dependent on fossil fuels, which are unsustainable in terms of supply and environmental impact.

Despite these enormous challenges, 'it is estimated that 30–50% (or 1.2–2 billion tonnes) of all food produced on the planet is lost before reaching a human stomach' (Institute of Mechanical Engineers, 2013: 2)! This is even more shocking given vast inequalities in food supply. Smith (2019) notes the following:

- 95% of the world's hungry live in the Global South;
- in some countries, 70% of adults are obese or overweight;
- women provide 40–80% of agricultural labour but only own 10–20% of land, and they receive only 7% of investments;
- 40% of global grain is used to feed livestock (enough for 3.5 billion people);
- the agribusinesses Monsanto, DuPont and Syngenta control half of the world's commercial seed and pesticide supplies.

Given these concerns, agriculture has rightly been a long-standing and central component of rural geography (Coppock, 1971; Grigg, 1984; Robinson, 2004; Maye, 2016), and has been studied across a range of scales, from the microscopic (Hinchliffe et al., 2013) to the global (Lawrence, 2019).

This chapter considers how geographers have studied farming by outlining three food regimes. A regime refers to a set of 'geopolitics, business economics, consumption patterns, retailing, processing, farming and supply industries' (Robinson, 2004: 43) that support a particular way of producing and distributing food. Although questions have been raised about the extent to which regimes have changed (Evans et al., 2002), the consideration of farming regimes is helpful because it considers how political and economic structures support and shape how food is grown (Table 5.2) (Maye, 2016; Robinson, 2004).

5.2 Productivism: Grow More Food!

The obvious solution to food shortage is to grow more food. This simple, perhaps overly simple, approach was widely adopted in response to shortages and threats to the supply of food in the Second World War, as well as failings in the previous colonial regime (Table 5.3) (Maye, 2016; Robinson, 2004).

TABLE 5.2 Agricultural Food Regimes

	Colonial	**Productive**	**Post-Productive**
Products	Grain, meat	Grain, meat, durable food	Fresh, organic, reconstituted
Period	1870s–1920s	1920s–1980s	1990s–present
Capital	Extensive	Intensive	Flexible
Food systems	Exports from family farms in settler colonies	Transnational restructuring of agriculture to supply mass markets	Global restructuring, with financial markets linking production and consumption
Characteristics	Culmination of colonial organisation of pre-capitalist regimes; the rise of nation-states	Decolonisation; consumerism; growth of forward and backward linkages from agriculture	Globalisation of production and consumption; disintegration of national agro-food capital and state regulation; green

Source: Robinson (2004)

TABLE 5.3 Agricultural Transformations

	Traditional	**Modernising**	**Industrialised**
Share of agriculture in GDP	High	Medium	Low
Consumption	Rising calorie intake, diversification of diets	Diversification of diets, move to processed foods	Higher value, processed foods
Retail	Small-scale and wet markets	Diversification of retail, spread of supermarkets	Supermarkets are widespread
Processing	Limited	Growing opportunities	Large processing sector
Wholesale	Traditional with a domestic focus	Traditional and specialised	Specialised; retailers often bypass these via their own distribution centres
Procurement	From local farmers, small market trading	From local and national markets, markets are regulated	From national and international chains, via managed chains; advanced arrangement
Production systems	Diversified low-input	Semi-intensive farming; a mix of family farms and corporate units	Specialised, intensive, 'productivist' farming, large family farms and corporate farms
Safety in food systems	No traceability	Traceability in some chains via private standards	Global GAP, hazard analysis and critical control points, private standards
Vertical coordination	Relationships	Relationships and rules	Binding agreements
Examples	Bhutan, Cambodia, Kenya, Laos	Brazil, China, India, Mexico	Australia, Canada, USA, EU countries

Source: Lawrence (2019)

In the 19th century, a colonial food regime saw settler states, such as Australia and Canada, export meat and grain to Europe and, by return, import manufactured goods from a rapidly urbanising Europe (Friedmann and McMichael, 1989). Yet these relationships were far from equal and relied on the invasion and exploitation of indigenous land and people for European markets. In Australia, for example, colonisation led to the decimation of indigenous people through disease and the destruction of traditional nomadic practices. The introduction of livestock – initially from the UK but, later, from other parts of the world (Tonts et al., 2010) – was particularly egregious, as it prevented traditional hunting practices and displaced indigenous people to marginal lands and settlements.

The culmination of colonialism, the rise of newly independent nation-states and a crisis of European food supply in both World Wars, drove a transition away from colonial systems to those of 'global capitalism and the pursuit of lower cost and more flexible sourcing of food' (Rosin, 2013: 53). Building on systems of measurement and standardisation established under colonialisation, this transition was supported by new forms of global finance (Gertel and Sippel, 2019: 215). Fordism – or the mass production of goods and methods – introduced methods of industrial production and consumption to farming (Cloke and Goodwin, 1992). For example, as farmers in the Midwest of the United States invested in tractors and machinery, their farms not only became more productive but also became the bulwarks of consumerism that supported the mass production of these goods (Page and Walker, 1991).

These circumstances led to the development of the productivist food regime (Table 5.2), which was characterised by a series of state policies and capital investments that sought to drive significant growth in food production. These trends accelerated after the Second World War when governments in many Western European countries pursued policies aimed at ensuring a safe, plentiful and uninterrupted supply of cheap food for their populations. In the UK, for example, the 1947 Agriculture Act was a piece of watershed legislation that guaranteed prices for farmers, incentivising them to grow as much food as possible in the knowledge there would always be a market for their produce. This transition was supported by new agencies that undertook research and provided advice to farmers, as well as food marketing boards that encouraged the public to consume this agricultural produce. Productivist farming was supported by the 1947 Town and Country Planning Act, which affirmed that rural areas should be preserved for agriculture at the expense of urban growth or social development (Curry and Owen, 2009).

Similarly, in 1962, the introduction of the Common Agricultural Policy (CAP) by the European Economic Community (EEC) aimed to use a system of farming supports and guaranteed prices to:

- increase production through optimum utilisation of resources;
- provide a fair standard of living for farmers;
- stabilise agricultural markets;
- guarantee food supplies;
- ensure reasonable food prices.

Agriculture became part of a globalised food system (Figure 5.2) that strengthened vertical connections between agrotechnological industries, farms and retail networks (Murdoch, 2000). As a consequence, productivist farming has been characterised by three main trends (Maye, 2016):

FIGURE 5.1 Productivist agriculture in the Western Australian Wheatbelt.

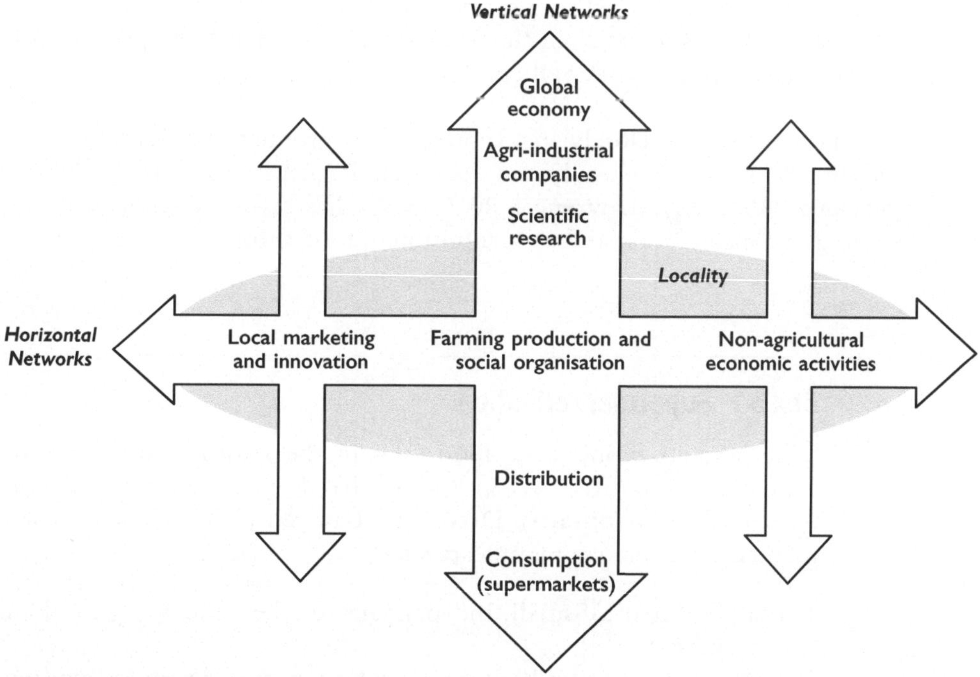

FIGURE 5.2 Agri-food networks.
Source: Murdoch (2000)

1. Concentration. Farming became concentrated into a smaller number of large farms, with a consequent decline of smaller and medium-sized family farms. This is best seen in regions of intensive farming, such as the American prairies or the Western Australian Wheatbelt (Figure 5.1). Currently, farms in Eastern Europe are undergoing amalgamation as investors seek to develop larger, more profitable enterprises.
2. Increase of capital inputs. The concentration of farming reflects growth in capital inputs, such as non-organic fertiliser, machinery or specialised breeds of livestock. Examples include breeding animals through very selective forms of artificial insemination (Holloway et al., 2009) and use of computer technologies, such as global positioning systems (GPS), to develop forms of precision farming that ensure the most profitable use of land and resources (Higgins et al., 2017; Visser et al., 2021). At its most extreme, technologies have been used to develop genetically modified foods (Dibden et al., 2013) and even artificial forms of meat (Tuomisto and Teixeira de Mattos, 2011). Given the high cost associated with these approaches, farmers have been prompted to concentrate on particular forms of farming in order to maximise economic returns on their investments. For example, given the cost of an automated milking parlour, it makes sense for farmers to concentrate on keeping dairy cattle rather than beef animals or growing crops.
3. A growth in the processing and manufacturing of food. There has been an increase in the consumption of processed and refined food that is high in saturated fats and sugar and low in fibre, which has contributed to a rise in obesity and associated health problems. The processing and distribution of food has become closely linked to a growth in supermarkets (Box 5.1), which has also distanced consumers from the production of food and the practices of farming (Boogaard et al., 2010; Sellick, 2020).

Supermarkets are closely incorporated into productivist farming systems, providing food relatively cheaply and delivering it 'just in time' using global transport networks (Box 5.1) (Lawrence, 2016). This has favoured large-scale producers, leading to consolidation and concentration of food through key suppliers and firms.

Box 5.1 Supermarketisation

Supermarkets monopolise food sales in the Global North and are also growing in significance in the Global South (where they currently account for 10% of sales). Dixon and Banwell (2019) identify that supermarketisation has six main features:

1. displacement of small, independent retailers, including local and wet markets;
2. dominance of purchasing power by supermarkets by providing low-cost, quality-assured products;

3. diversification of services beyond food and hardware retailing to position themselves as trusted authorities, with the growth of online delivery representing a further expansion;
4. development of regulation and supply chains, leading to direct forms of food processing – through their own facilities – or indirectly through contracts;
5. development of own brands, which now account for their grocery sales;
6. global supply chains to import fresh and processed products through 'just in time' delivery systems.

Supermarkets have been able to exert greater control over farmers and food manufacturers through contracts and supply chains. This has put pressure on smaller producers and encouraged larger farms to invest in technology to meet the quality standards of supermarkets. Some have suggested that the private sector rather than the government is now the main regulator in rural places, something that is contributing to the wider marginalisation of rural communities. As some campaigners have pointed out, exacting requirements to meet consumer expectations have led to food waste, often on cosmetic grounds. There is also a certain irony that large supermarkets are often absent from agricultural areas, given low population densities, meaning that food is more expensive in many rural places. Further, a desire to meet consumer demand for fresh goods all year round contributed to a fourfold increase in food exports between 1961 and 1999 (Maye, 2016), as well as increases in food miles (Shaw, 2014).

The importance of supermarkets in the supply of food and household goods was underlined during the Covid-19 pandemic when panic buying emptied shelves of some products, such as pasta, long-life milk, fruit and toilet paper. This was not caused by a shortage in food supplies but, rather, a lack of warehoused stock and 'just in time' delivery systems that were unable to cope with unforeseen demands. As the pandemic progressed, supermarkets were required to ration products, prioritise customers most in need and review their range of products, in some cases specialising to speed up delivery. It was supermarkets, rather than farmers, that were involved in strategic decision-making about food distribution, highlighting their central place in the strategic delivery of food in developed countries.

As the global networks of farming have expanded, food and land have become important financial commodities (Gertel and Sippel, 2019), reflected in land grabs (Chapter 5) and speculative investments in food. Thus, in 2008, wheat prices reached a record high of $12 a bushel (including a 25% increase on 25 February 2008) (Figure 5.3). In part, this reflected poor harvests and export restrictions and, more significantly, investment by traders who regarded wheat as a safe investment during a period of global recession and falling oil prices (BBC 2008). This

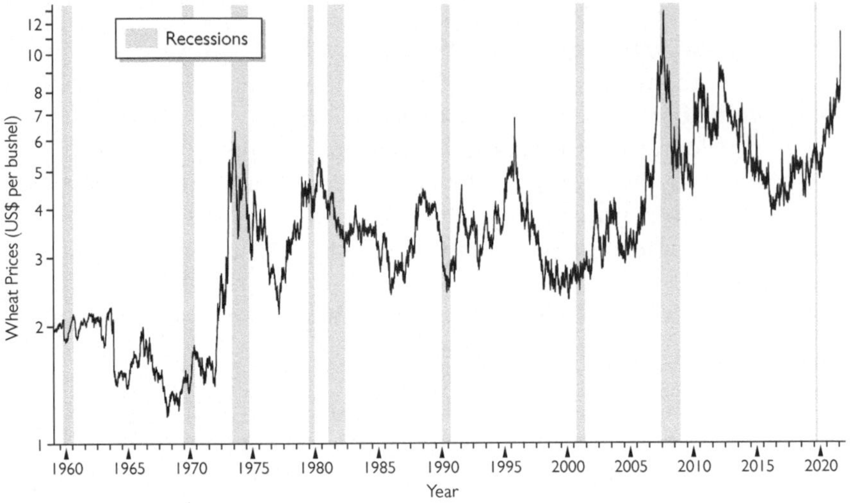

FIGURE 5.3 Global wheat prices.
Source: Macrotrends (2023)

inflation led to political unrest in some parts of the world and contributed to the Arab Spring uprisings. At the time of writing, the invasion of Ukraine has pushed wheat prices to over $12 a bushel and, globally, rising food prices are contributing to high rates of inflation. These increases have been driven by restrictions on food exports as well as higher production costs due to growing energy and transport prices. These challenges highlight the global, economic and social significance of food production as well as the importance and fragility of international supply chains in the delivering of produce.

In recent years, a growth in demand for biofuels has also led to speculative investment in agricultural land, with implications for food security (Food and Agriculture Organization of the United Nations (FAO), 2009). As Gertel and Sippel (2019: 222) note, 'livelihoods in the global north (seeking long term gains by pension schemes) and livelihoods in the global south (the risk of food insecurity) are connected by speculative financial activities . . . shareholder value has often been realised at the expense of the rights of suppliers, workers and customers' (220).

Box 5.2 Changing Farming Policies: The Example of the European Union

The Common Agricultural Policy (CAP) was introduced in 1962 by the European Economic Community (later known as the European Union) as a price support policy for farmers. CAP aimed to increase agricultural productivity through fixed prices and subsidies to encourage farmers to produce cheap food; ideals that reflected and drove the productivist era in farming. As the name suggests, the Common *Agricultural* Policy was targeted at a particular economic sector (agriculture) rather than the countryside as a whole; consequently it neglected other economic and social activities in the country (Lapping and Scott, 2019)

CAP led to highly intensive farming characterised by high costs, over-production, waste, environmental damage and the net export of food (Ackrill, 2000). An excess of dairy products led to quotas being imposed in 1984, but exports of cereals, beef and sugar continued to rise dramatically (Ackrill, 2008). As these foodstuffs were heavily subsidised, tensions arose with other food-exporting nations. In 2002, for example, sugar was subsidised at €400 a tonne, leading to the EU becoming the world's leading exporter of sugar. One and a half million tonnes were eligible for export refunds, 'at over twice the world market price for sugar!' (Gibb, 2004: 571). Producers elsewhere in the world found it difficult to compete with EU farmers because of these payments and pressure was put on the EU to reduce these subsidies.

The cumulation of these issues led to reforms of EU rural policy (Terluin, 2003). In 1992 the MacSharry Reforms saw a reduction in price support, particularly in cereals and beef production, and the introduction of single farm payments together with payments to 'set aside' land from agricultural production, plant trees and encourage early retirement. These measures indicate a shift towards post-productive farming and appear to have contributed to greater stability in agricultural spending (Ackrill, 2000).

Further reforms in 1999, known as Agenda 2000, established rural development as a 'second pillar' of CAP and doubled the budget (from around 5% to 10%) for this task (Ackrill, 2008). CAP continues to support farming but aims to do so in a more sustainable manner (Lapping and Scott, 2019; Vergunst, 2016), while promoting social inclusion, reducing poverty and encouraging economic development in rural areas. Subsidies were replaced by direct farm payments, which, as the name implies, were paid directly to farmers on a per-hectare basis and in accordance with environmental, food safety and welfare 'cross-compliance criteria'. Despite these reforms, larger, more affluent farms are more likely to benefit from CAP. Furthermore, the majority of EU rural funding continues to be centred on agriculture (Ackrill, 2008). and the majority of EU rural funding remains centred on agriculture (Ackrill, 2008).

5.3 Post-Industrial Farming?

Industrial farming increased production, providing cheap and plentiful food for large numbers of people, which was, after all, the aim. Yet increasingly, questions were raised about the budgetary costs of productivist policies, the waste caused by overproduction, health issues connected to processed food and environmental damage caused by intensive farming. The drive to productivity meant that vulnerable habitats were drained, ploughed over or pulled up in order to farm previously uncultivated land in ways that suited mechanised, intensive production. Rachel Carson's influential book

Silent Spring (Carson, 1962) graphically highlighted the potentially destructive use of pesticides. Since then, further concerns have been raised about the use of non-organic fertilisers and genetically modified crops. Diseases such as BSE, foot-and-mouth and avian flu impacted some farming sectors very heavily and were seen by some to be a consequence of intensive farming (Whatmore, 1997). For example, the outbreak of bovine spongiform encephalopathy (BSE; commonly referred to as 'mad cow disease') in the UK in the 1990s was thought to have been caused by meat-and-bone (MBM) meal that contained the remnants of diseased cattle, as well as scrapie-infected sheep, that, in turn, infected cattle that ate it (Hinchliffe, 2001). The subsequent outbreak of BSE led to an economically damaging ban on the export of British beef and, more widely, prompted ethical concerns about animal welfare (Buller and Morris, 2003).

Productivism also led to cracks in the social structures that supported farming. In places where industrialised farming grew, there was a decline in family farms and the number of people working in farming. This initially led to depopulation, the gentrification of former agricultural properties and, later, a more sustained trend of counterurbanisation (Newby, 1979b). State policies to diversify employment contributed to a decentralisation of manufacturing from urban to rural places and a growth in the service sector (Healey and Ilbery, 1985; Cloke and Goodwin, 1992). Some of these activities further challenged the significance of farming in rural places and reflected the class and property interests of new rural populations (Woods, 2006).

As a result of these pressures, there was a significant change in farming policy in the mid-1980s (Box 5.2). Some commentators considered that these policies marked the start of a new 'post-productive' agricultural regime characterised by extensive, less concentrated and more diverse or pluriactive farming activities (Ilbery and Bowler, 1998) (Table 5.5). Horizontal, rather than vertical, linkages (Figure 5.2) were re-emphasised as efforts were made to reconnect farms with other social and economic activities in their localities. While the extent of this transition has been debated (Evans et al., 2002), the aim was to counter or reverse many of the issues associated with industrialised farming. Three examples are listed here:

1. Landscape or environmental stewardship. A series of policy measures encouraged farmers to tend their land to benefit the environment or landscape. Initially, specific schemes targeted particular aspects of farming, such as setting aside land from production (Morris and Young, 1997), but over time, these became incorporated into wider agricultural policies, such as the EU's MacSharry reforms, that promoted more environmentally friendly forms of farming. Landowners have also been persuaded or required to open up land for public access, either through payments or changes in the law (Parker, 2006).
2. Diversification. Farmers diversified their on- and off-farm activities to expand their income streams. Pluriactivity, as it is known, includes the use of farm premises to support alternative economic activities, particular in the leisure and tourism sector. Examples include accommodation or using farm premises to host new enterprises, such as craft industries. These range from fairly minor modifications, such as using a spare room for Airbnb, to whole-scale changes to the operation of a farm. One example has been the establishment of farm parks as tourist attractions.

TABLE 5.4 Alternative Food Networks

Conventional	**Alternative**
Modern	Post-modern
Manufactured/processed	Natural/fresh
Mass (large-scale)	Craft/artisanal
Long food chains	Short food chains
Cost externalised	Cost internalised
Rationalised	Traditional
Standardised	Difference/diversity
Intensive	Extensive
Monoculture	Biodiversity
Homogenous	Regional palates
Hypermarkets	Local markets
Agrochemicals	Organic/sustainable
Non-renewable	Reusable energy
Fast food	Slow food
Quantity	Quality
Disembedded	Embedded
Source: Maye (2016)	

3. Alternative food networks. Maye (2016) identifies the emergence of new or alternative food networks (Table 5.4) that are characterised by the following:

a. Shorter food chains aimed at reducing food miles and offering opportunities to buy food directly from farmers (Figure 5.4). Vente direct schemes and farmers' markets also offer opportunities for consumers to engage more closely with the producers of food (Winter, 2003).
b. An emphasis on quality by, for example, using organic methods of production (Ilbery et al., 1999), rare breeds (Yarwood and Evans, 1999) or traditional/craft methods. These concerns reflect and support wider consumer trends such as 'slow' food or artisan bakeries, often supported by television cooks.
c. Ethical consumption, whereby consumers consider the political consequences of their purchases (Goodman et al., 2010). Fair trade produce, for example, reflects greater concern for the economic and social welfare of food producers; a switch to plant-based diets may be prompted by concerns for animal welfare, and the choice to purchase local or organic food may assuage fears about the environmental costs of food production (Mouat and Prince, 2018; Miele and Lever, 2013).

Of course, the ideas presented in Table 5.4 are binaries and, as such, mask a more complex reality (Winter, 2003). Thus, Morris and Buller (2003) suggest that the idea of 'local' food may reflect different, contested versions of localism:

1. Parochial or defensive localism, which supports local farmers and farming traditions (Winter, 2003). Particular produce, for example, might be used in place-marketing campaigns because of its association with specific localities (Hill, 2020).

FIGURE 5.4 Farmers market in Margaret River, Western Australia.

2. Flexible localism that uses localism tactically. The term 'local' might be used to describe a range of geographical scales, from 'within 5 miles' to 'British' or 'Western Australian'. Watts et al. (2005) also note differences between 'local suppliers', who draw upon local products, and 'local companies', who market their location but draw on produce from further afield.
3. Competitive localism, where different suppliers or retail spaces compete with each other for customers. For example, supermarkets with local produce lines may impact upon farmers' markets or other forms of direct sales.

There has also been considerable debate about the extent to which there has been a transition from productivist to post-productive farming (Evans et al., 2002; Mather, 2006: 720; Wilson, 2010; Walford, 2003). Evans et al. (2002) systematically question the extent to which farming has adopted features of post-productivism. Thus, quality food can also apply to mass markets, and indeed, many supermarkets have a high quality brand (e.g., Tesco Finest). Organic food accounts for a very small minority of production and there is also evidence that organic production is falling (Department for Environment, Food and Rural Affairs, 2018). While organic farming has a lower environmental impact *per acre of land* than conventional farming, Tuomisto et al. (2012) demonstrate that the environmental impact *per unit of food* is higher given lower yields. It has also been suggested diversification is linked to existing activities and that extensification, such as set-aside, is applied to the least productive areas of land (Evans et al., 2002). Farming remains so resolutely intensive that in some regions, such as East Anglia or the Mid-West prairies, it has been described as 'super-productive' (Halfacree, 2007). While post-productivism can be applied to understand some aspects of farming change in Europe, it does not account for change in other parts of the world, especially the USA and Australia (Argent, 2002; Lawrence, 2019).

Neo-productivism

Despite apparent shifts in farming, there is continuing evidence of 'neo-productivism' or the continuation of productivist farming in modified or subtle ways (Wilson and Burton, 2015). Evans (2013) surmises that neo-productivism is characterised in four ways.

1. Environmental awareness. Farming continues in a high-output fashion but attempts to do so in more environmentally friendly ways (Rannikko and Salmi, 2018). Examples include reforms to the Common Agricultural Policy (Box 5.2) that have attempted to enrol ideas of sustainability into what essentially remain as productivist farming systems. There have also been 'technocentric' (Robinson, 2009) efforts to farm more sustainably, such as using different breeds of livestock or reducing water consumption (Chapter 13).
2. Farming as a meaningful occupation – most farmers continue to regard farming as an activity that should focus on the production of food rather than the curation of landscapes. Hence, rewilding may seem at odds with farming traditions that have emphasised the management of land for food and may have encapsulated efforts that, over many generations, made land productive or useful for farming.
3. How agriculture is organised. There is a dualism between heavily capitalised farming and smaller, subsistence farms (Robinson, 2018), which is starkly illustrated in the EU. Thus, nearly one-third of family farms are found in Romania, which contrasts with larger, industrialised farms in Western Europe. Globally, 90% of farms are run by families, which occupy 70–80% of farmland and produce 80% of the world's food (Brunori and Bartolini, 2019). Yet despite this apparent diversity, supermarketisation, industrialisation and reorganisation continue to pressure family farms towards agglomeration and industrialisation. Evans (2013: 63), for example, draws attention to the rapid growth of plastic poly-tunnels by smallholders in the UK as 'a horticultural innovation tied to the power and preferences of supermarket food retailers'.
4. Regulatory systems. Efforts to regulate farming through planning regulations and policy frameworks have only been partially successful. Woods (2006) has argued that rural politics has shifted away from agriculture and towards a much wider set of rural interests that, in some cases, have challenged farming activities and infrastructure. Despite opposition to some agricultural activities, such as intensive 'megafarms' or the use of poly-tunnels, these are often able to 'outflank' existing regulations, in part due to their novelty, to establish new forms of intensive production (Evans, 2013).

Additionally, some farmers have adopted productivist principles to grow new, non-food crops. Biofuel production increased more than threefold between 2000 and 2008 (Food and Agriculture Organization of the United Nations (FAO), 2009) and, by 2019, had reached 161 billion litres. Concern has been expressed that the use of grains for biofuel production will impact food security (Subramaniam et al., 2019). One study estimates that continued increases in biofuel production will increase the number of undernourished preschool children in Africa and South Asia by up to three million (Food and Agriculture Organization of the United Nations (FAO), 2009). Investment in crops for biofuels has also impacted food prices, with one study attributing 20–40% of the 2008 food price rises to biofuel trading (Mittal, 2009). While biofuels may contribute to environmental sustainability, questions remain about the economic and social impact of these new forms of production.

Box 5.3 Agri*cultural* Geographies

Addressing the Geographical Association in 1959, Ogilvie Buchanan (1959: 2) noted the importance of social institutions and customs in farming, commenting that 'the geography of agriculture is properly an exercise in social geography' and that the 'complete conversion of farming to purely economic fancies is still far from universal'. An academic focus on changing farming regimes has led, perhaps, to a neglect of farmers as individuals. Farming is also a way of life, which embraces particular ways of behaving, working and living. Brian Ilbery (1978) pioneered the use of behavioural geography to appreciate how farmers' perceptions of land use and agricultural practices could be used to understand farming change. This approach emphasised the agency of individual farmers and decision

FIGURE 5.5 Judging livestock at an agricultural show.

making, such as examining who were 'laggards', innovators or early adopters of particular farming practices.

Inspired by the cultural turn, there has also been a call to study agri-*cultural* geography or the ways in which farming is culturally represented and reproduced through different practices (Morris and Evans, 2004). This recognises that farming is more than an economic activity but is also a social activity that enrols animals, landscapes and other farmers. For example, the physical attractiveness of an animal or the aesthetics of a field of crops reflects what is regarded as 'good', or conversely bad, farming practice (Burton, 2004). In turn, these ideas are maintained through institutions such as breed societies (Yarwood and Evans, 2006), agricultural shows (Holloway, 2004) and daily practices of farming (Gray, 2014) (Figure 5.5). Family structures and gendered expectations of men, women and children remain very important in farming (Riley, 2009). Attention has also been given to the way that farming is presented to non-farming audiences – something that is of growing importance given debates about animal welfare and sometimes militant responses to livestock farming (Ceccato, 2022).

5.4 Forestry

Commercial forestry has mirrored many of the changes that have occurred in agriculture. In some localities, it remains a significant economic activity that draws on productivist principles to supply timber and wood pulp for construction and paper manufacture. It has been estimated that forestry employs 13.2 million people globally (FAO, 2014), although Lippe et al. (2021) estimate that between 64 and 94 million people are employed formally or informally, directly and indirectly in forestry-related industries, especially in India (10–12% of the workforce), China, Brazil, Ethiopia, DR Congo and Indonesia. In the Global North, forestry is a significant activity in North America and Scandinavia. In these places, timber production is owned by multinational companies and responds to demand from global markets (Halseth, 2018). At the time of writing, rising timber prices are being driven by high global demand from the construction industry as it emerges from Covid-19 lockdowns.

Forestry, like food production, has been valued as a strategic resource and, as such, has received government support and regulation. Forests have been managed on productivist lines with:

> *ever increasing levels of production, automation, and capitalization; growing corporate dominance; a dramatic disembedding of commodified forest products from their local social and environmental contexts as they were propelled into increasingly global markets; and diminishing returns to forest-dependent communities.*
>
> *(McCarthy, 2006b: 806)*

These approaches have been challenged by growing environmental concerns about logging, loss of biodiversity and the release of carbon (Halseth, 2018). Consequently, there have been efforts to restructure and re-evaluate the significance of trees and forestry to society. In some areas, there has been a growth in community forests that are locally managed to support forms of ecological, economic and social sustainability (McCarthy, 2006a). It has been suggested that while community forests can have positive environmental and economic benefits, access and resource rights are more likely to be contested at the local level (Hajjar et al., 2021). In one case in Senegal, senior men used the governance of the community forest to assert their patriarchal authority, while others used their indigeneity to emphasise their role as environmental stewards (Robinson, 2021). Consequently, women and people from other ethnic groups (known as strangers or 'second comers') find it harder to access forest resources or participate in their governance. Although these forms of post-productive forestry offer a counter to productive, top-down approaches to managing trees, they reflect neoliberal forms of governance that shift the responsibilities of forest management from government to community (McCarthy, 2006b). Community, as Chapter 10 explores, is not a panacea for rural problems and brings with it new, or rather existing, imbalances of power that continue to exclude certain groups. It is, therefore, important to consider the relationship between economic and social change in different rural settings.

5.5 Conclusions

Productivism and post-productivism are not, and were never posited to be, simple binaries. Geography matters (Massey and Allen, 1984), and we should expect to see different elements of productivism and post-productivism in different places. Halfacree (2007) argues that spaces of productivism or even super-productivism continue to be found in rural places, together with more alternative visions of rural life. Transition to industrial or post-industrial farming is not inevitable, prescribe or even desirable, with commentators recognising that there are many different pathways that reflect the outcomes of different social, economic, political and environmental processes (Wilson, 2001) (Figure 5.6). The term 'multifunctional' is used to reflect that different farming systems may operate simultaneously, reflecting a spectrum, rather than a binary, of post-productivist/productivist practices. Further, regimes may be unstable and challenged by various shocks that may propel farmers and farming into new ways of producing food (Maye, 2016). Indeed, the current climate emergency is prompting new transitions in farming regimes in order to provide food securely, safely and sustainably (Chapter 13).

TABLE 5.5 Features of the Post-Productive Countryside

	Productivist Countryside	**Post-Productivist Countryside**
Dominant Economic Activity	Agriculture	Service sector
Mode of Production	Production	Consumption
Mode of Regulation	Fordist	Post-Fordist
Farming Characteristics	Intensive Concentrated Specialised	Extensive Dispersed Diverse
Food	Quantity	Quality
Politics	Rural politics	Politics of the rural

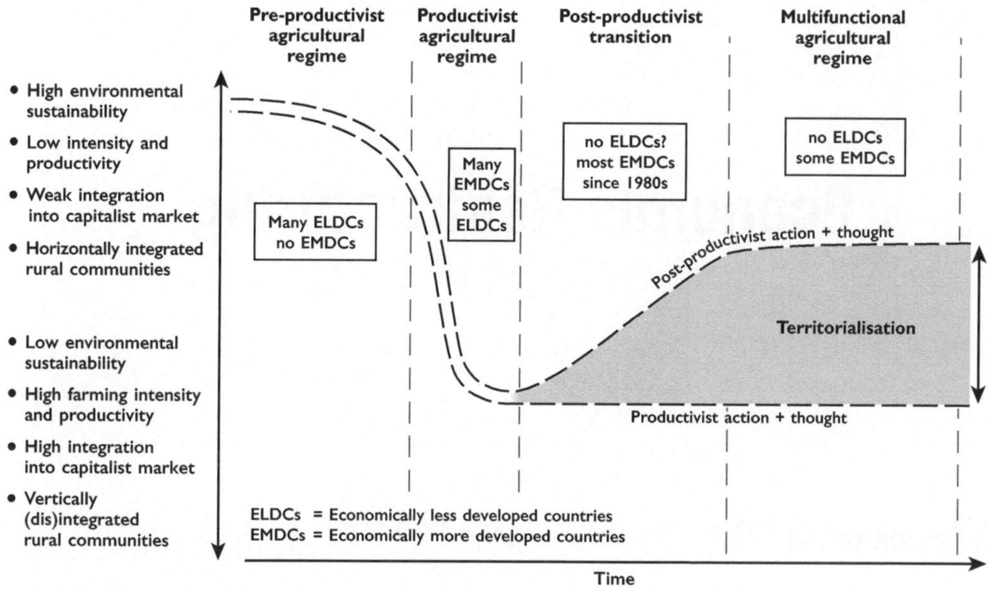

FIGURE 5.6 Agricultural regimes.
Source: Wilson (2001)

Finally, the post-productive countryside is not only used to refer to changes in farming practices (Atterton, 2016). It is also used to describe a countryside in which the production of food is no longer the dominant industry (Table 5.5). As the next chapter shows, other economic sectors, including manufacturing and tourism, have grown in economic and social significance (Table 5.5).

CHAPTER

6

Economic Restructuring

6.1 Introduction

The previous chapter outlined some of the key changes that have occurred in agriculture. In addition to farming, these have also had three important consequences for rural economy and society.

First, while agriculture still accounts for a significant proportion of land use, the number of people working in this sector has declined (Figure 6.1). Consequently, rural society is no longer centred on farming and, instead, has become economically, socially and culturally diverse. Second, as a consequence of agricultural restructuring there have been opportunities for new forms of economic investment. For example, manufacturing industries have grown in rural places, partly to take

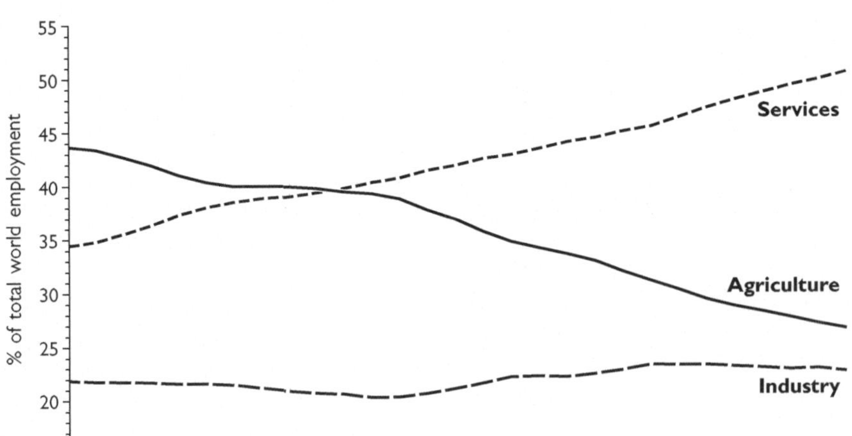

FIGURE 6.1 Global agricultural employment.
Source: The World Bank (2023)

DOI: 10.4324/9780429448966-8

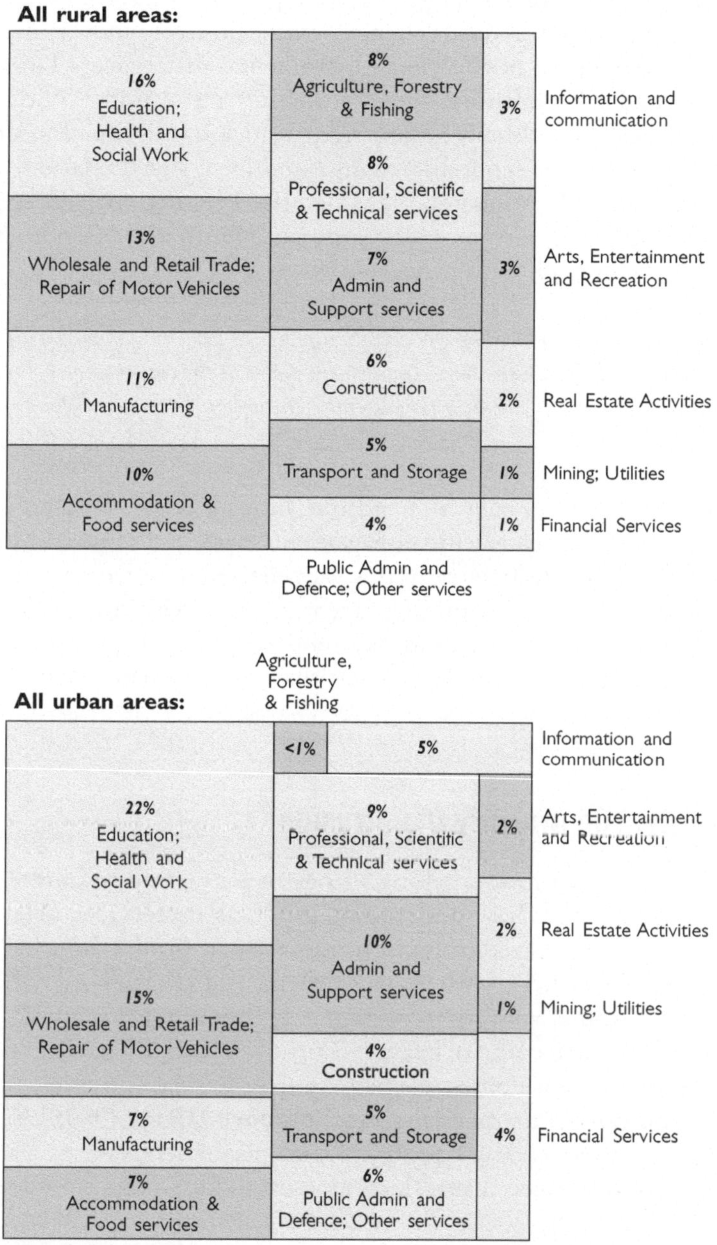

FIGURE 6.2 Employment by sector in rural and urban England, 2019/2020.
Source: (DEFRA, 2021)

advantage of plentiful, cheap labour as agricultural employment has declined (Healey and Ilbery, 1985; Cloke and Goodwin, 1992). This has happened to such an extent that manufacturing employs relatively more people in rural areas than in urban ones (Figure 6.2)! Finally, the globalisation of farming has connected the countryside more closely to international networks of production, investment, transportation and consumption (Robinson, 2018; Woods, 2007).

It is also important to note that change is geographically uneven, with some localities undergoing rapid and radical change, with positive and/or negative impacts, and others experiencing negligible or incremental differences. Thus, primary industries associated with farming, forestry, fishing and mining continue to be important in some localities, although their significance has declined in others. Murdoch et al. (2003) trace how economic change and local social relations combined to form a 'differentiated countryside' characterised by regional differences, diverse identities and differing forms of development. Similarly, Woods (2007) recognises that globalisation is a hybrid process in which:

> *the networks, flows and actors introduced by globalization processes fuse and combine with extant local entities to produce new hybrid formations. In this way, places in the emergent global countryside retain their local distinctiveness, but they are also different to how they were before.*
>
> *(Woods, 2007: 500)*

In this sense, globalisation is never fully attained but, instead, occurs to a greater or lesser extent depending on 'locally specific engagements with and responses to globalization' (Woods, 2007: 486). Thought of in this way, economic restructuring might be best seen as a process rather than a finished event. And so, with the qualification that restructuring impacts different places in different ways, this chapter goes on to outline some of the significant economic changes that are occurring in the global countryside.

6.2 Consuming Rurality: The Amenity Countryside

If production refers to growing food, making things or extracting resources, consumption refers to the sale, purchase and use of products or services (Mansvelt, 2005). And in many rural places, consumption, rather than production, has been the main driver of rural change. Very often, this reflects and relies upon marketing and selling a particular vision of rurality to consumers. Tourism, for example, relies on visitors 'buying into' a particular, packaged vision of rurality by paying to enter particular sites (such as a stately home), participating in leisure activities, renting accommodation, buying souvenirs or eating local produce (Urry, 1995; Urry and Larsen, 2011). Buying a house in the country, either permanently or as a second home, is also a consumption choice that reflects a desire to buy into a rural location or lifestyle.

The idea of the 'amenity countryside' has been used to analyse how rural places have been transformed by and for consumption. Amenity value is the perceived attractiveness of a rural area and how its qualities and facilities attract and support rural lifestyles (Argent et al., 2014). It can refer to physical characteristics, particularly lakes, rivers, woods, national parks, beaches or vistas; and distinctive or differentiated cultures, expressed through art, language, food, wine or language; and greater opportunities for leisure, learning and, sometimes, spirituality (Matarrita-Cascante et al., 2015; Perkins et al., 2015) (Figure 6.3). In an Australian context, Argent et al. (2007) identify a series of environmental, locational and socioeconomic criteria (Table 6.1) – including annual rainfall, terrain and altitude, remoteness, location of

FIGURE 6.3 The amenity countryside of Noss Mayo in Devon.

settlement, irrigation water resources, distance from a beach, and employment in recreation and related services – as significant contributors to amenity value. These qualities are found in accessible, temperate, seaboard locations that have, consequently, attracted high numbers of in-migrants seeking rural or coastal lifestyles (Argent et al., 2007).

TABLE 6.1 Facilities Contributing to Amenity Value

Timescale	Activities	Facilities Contributing to Amenity Value
Immediately	Travel between places	A bus shelter, public toilet or roadside emergency telephone
Daily	Daily life by permanent residents	Medical facilities, the perceived congeniality and safety of the social environment, along with the possibility of pursuing favourite pastimes such as golf, fishing or skiing.
Diurnally	Overnight stays as part of longer journeys	Choice of stop-off places on a journey, including the availability and quality of overnight accommodation, eating places, barbecue facilities and swimming pools
Weekly	Day visits or weekend trips to towns by rural residents or visitors	Perceived quality of sporting, entertainment, and shopping facilities; social clubs or community buildings
Annual	Holidays or business trips	The weather, scenery, beach access, tourist attractions and the night-life experienced on holidays or longer trips

Source: Argent et al. (2007)

Amenity value relies on the exploitation and commodification of natural resources, which can be amplified through the designation of national parks, heritage sites and other forms of protected area status (see Chapter 14). Queenstown in New Zealand, for example, has acquired high amenity value through adrenaline sports (Cloke and Perkins, 1998), high quality rural living and spectacular mountain scenery (Perkins et al., 2015). Consequently, Queenstown has attracted international in-migration, overseas investment in property and increasing numbers of global tourists, reinforcing its amenity value at a global scale (Woods, 2011a). In turn, the social and natural landscapes of Queenstown are being developed to meet and maintain amenity aspirations of international visitors. Local people remain significant in supporting, opposing or partnering these developments (Woods (2011a).

Amenity landscapes exist alongside farming rather than replacing it (Argent et al., 2014). Thus, agricultural activities can add to the attractiveness of amenity landscapes: irrigation, for example, is essential for farming in arid areas but also makes these places attractive to in-migrants. Some in-migrants engage in lifestyle or hobby farming, diversifying farming practices and enrolling them into farming networks (Tonts et al., 2010). Productivist agriculture can therefore still occur in amenity landscapes albeit in smaller parcels of land (Argent et al., 2014).

Many regions with high amenity value have experienced population growth and an apparent surge in economic fortunes (Saint Onge et al., 2007; Hunter et al., 2020). Yet a growth in tourism and recreation industries relies on a low-waged workforce who are employed in low-skilled, insecure, part-time jobs (Shucksmith and Brown, 2018). Consequently, amenity-based economies are often characterised by social inequalities between wealthy migrants able to invest and live in amenity landscapes and more transient populations working in leisure-based employment who cannot afford to live permanently in these places (Hunter et al., 2005).

Crucially, the development of amenity value also links the countryside to new networks of global trade (Box 6.1). In the Global South, the purchase of land to develop new amenities for migrants or tourists is often linked to international investments or land grabs (Zoomers, 2010). These intrusions 'profoundly transform landscapes and are advanced by major capital investments . . . turning undercapitalised scenic areas into (prime) real estate locations' (Rainer, 2019, p. 1369). Indeed, over recent years, rural land and property have become sought-after international commodities, channelling 'new money' and new people into rural places (Gallent et al., 2018; Phillips and Smith, 2018a). Thus, migration from North to South America is driven by a promise of a rural idyll, cheaper lifestyles and investments in amenities, such as gated housing or leisure clubs with exclusive golf courses or marinas (Rainer, 2019). It also reflects the growing significance of international investment in rural areas.

Box 6.1 Locality Change in Cromwell District, New Zealand

Amenity value was central to the transformation of Cromwell District, Otago, New Zealand in the early 2000s (Perkins et al., 2015). The district had high amenity value due to its spectacular mountain scenery, associations with mining heritage and investment opportunities associated with the development of a dam and reservoir, all of which attracted tourists and in-migrants. Some migrants had associations with the wine industry and, through a process of 'creative enhancement', successfully established new vineyards. Subsequently, 'viticulture, wine-making and amenity residence' led to a rapid increase in population and investments in the wine industry (Perkins et al., 2015: 90). Between 1996 and 2013, grape harvesting increased from around 500 tonnes per annum to nearly 8,500, serving 124 wineries (compared to 11 in 1996) that produced two-thirds of the region's wine.

One resident described how wine-making changed from a cottage industry to an international enterprise:

> *Fifteen years ago there was no expertise [in the District]. The person who put our posts in had to develop, as a contractor he'd done some fencing and that sort of thing, but he had to develop his own tools to drive posts for us. The local engineering company had to make things for us too. For instance, we needed a machine to put on and lift [bird protection] nets. We'd go down there with a picture from a magazine and say: 'can you make this?' Things like even getting vines was not straightforward . . . anything specific to the industry. Then . . . if you did want something, the local stock and station people would know what was available for horticulture [pip and stone fruit production] but they wouldn't have a clue what to apply for grapes. So in a short space of time we've moved from that, which was absolutely pioneering stuff in terms of people providing goods and services to the industry, to a situation now where specialist companies have become well-established. They've put in depots, they've built here and they've developed the knowledge.*
>
> *(Perkins et al., 2015: 90)*

Although the region already had international trade links through the wool trade and tourism, Perkins et al. (2015) argue that amenity migration 'reassembled' the region to exploit its growing reputation for producing wine. Some amenity migrants from New Zealand and overseas invested in vineyards for prestige; others saw opportunities to support the wine industry using their marketing, hospitality or marketing skills. Alongside these investors, local farmers capitalised on the growth in viticulture by selling land for vineyards, growing grapes or diversifying into accommodation. Despite some limited opposition, these activities built on the region's amenity landscape to link it more closely with the global wine trade and place 'pinot next to merino in a spectacular high country landscape' (Perkins et al., 2015: 96).

6.3 International Investment

Land grabs, in which wealthier countries or investors buy or lease land overseas to cultivate crops or exploit resources, are the starkest form of international investment in rural places (McDonagh, 2014; Batterbury and Ndi, 2018; Nolte et al., 2016; Lay et al., 2021). Ostensibly, land grabs are driven by a desire to secure land for food or biofuel production or to offset carbon emissions through planting trees (Hunsberger et al., 2017). However, the rate of land-grabbing increased after the 2008 financial crisis when investors came to regard land as a tangible asset that was more likely to retain its value than stocks and shares (Pearce, 2012).

These actions reflect what Carmody and Ofori (2020) refer to as an 'ecological contradiction', whereby economic growth in the home country is restrained by a lack of resources. Investors with economic and political power overcome these constraints by buying land in other states. Land is regarded by investors as a commodity and a means to making profit but, to those who live on it, land represents a way of providing income or food. As such, some view land-grabbing as providing opportunities for supporting livelihoods and stimulating agricultural growth in less economically developed countries, while others are concerned with losses of land, the marginalisation of family farming and the stripping of assets to richer countries (Cotula, 2012). Indeed, land-grabbing occurs most often in areas with higher rates of hunger (Nolte et al., 2016).

The trend is 'truly global' (van der Ploeg et al., 2015): between 2012 and 2016, the Land Matrix Database recorded that the number of agricultural land deals doubled from 323 to 604, rising from 1.7 million hectares to 6.4 million hectares. The majority of these deals were for food crops (553 deals and 9.2 million hectares), followed by oil palm (263 deals on 5.6 million hectares) and agrofuels (221 deals on 5.1 million hectares) (Nolte et al., 2016).

Forty percent of land-grab deals occur in Africa, especially in Sudan, Ethiopia, Madagascar and Mozambique (Table 6.3) (Nolte et al., 2016). According to the World Bank, many of these transactions are made by citizens of the country in which they are purchased, although subsequently these are often transferred to international investors (Cotula, 2012). Many land-grabbing deals have occurred in Southeast Asia (Cambodia, Laos, Philippines, Indonesia) and South America, although Oceania accounts for the largest area of land sold in this way (Cotula, 2012). Land-grabbing is also significant in Eastern Europe. Over five million hectares of land were purchased in Russia and Ukraine in 2016, although the war in Ukraine has clearly curtailed this trend. EU has also recognised that land-grabbing is a 'creeping phenomena' in many of its Eastern European member states (Box 6.2) (Sylvia et al., 2015; Constantin et al., 2017). Although media interest has focused on the influence of China, Figure 6.4 reveals that Malaysia has been the largest investor country, followed by the USA and the UK.

Box 6.2 Land-grabbing in Eastern Europe

Although land-grabbing has widely been associated with countries in the Global South, it is also having a significant impact on Eastern Europe and Eurasia (van der Ploeg et al., 2015; Visser and Spoor, 2011). The Common Agricultural Policy (CAP) and state governments have favoured bigger farms, incentivising the purchase of land to support the concentration of farming. Cheap land prices (that were five to ten times lower than in Western Europe in 2015) have further driven the purchase of small, family-run holdings that were struggling to compete with industrial-scale farming.

Six percent of farmland in Romania is owned by transnational corporations (Bouniol, 2013). One agribusiness has invested more than €10 million in the Cluj District of Romania. As well as receiving EU subventions to modernise food processing, it has also bought out other commercial companies and rented villagers' small holdings extremely cheaply. In many countries, transnational deals have been aided by *arendatori*. Arendatori are financial agents who buy land cheaply and then rent or transfer it to transnational investors, thereby circumventing laws that prevent the purchase of land by overseas investors (van der Ploeg et al., 2015).

In Eurasia, concerns have been raised about 'de-peasantisation' or the loss of traditional rights, practices and social organisations associated with small-scale production. It is consequently feared that land-grabbing will lead to social inequalities and loss of local power. It is driving a shift to productivism that is characterised by lower levels of agricultural employment (Emiliana in west Romania only generates employment for 99 people in an area of 12,000 hectares (Bouniol, 2013)) greater specialisation and, consequently, the expansion of monoculture through greater concentration of production on large estates.

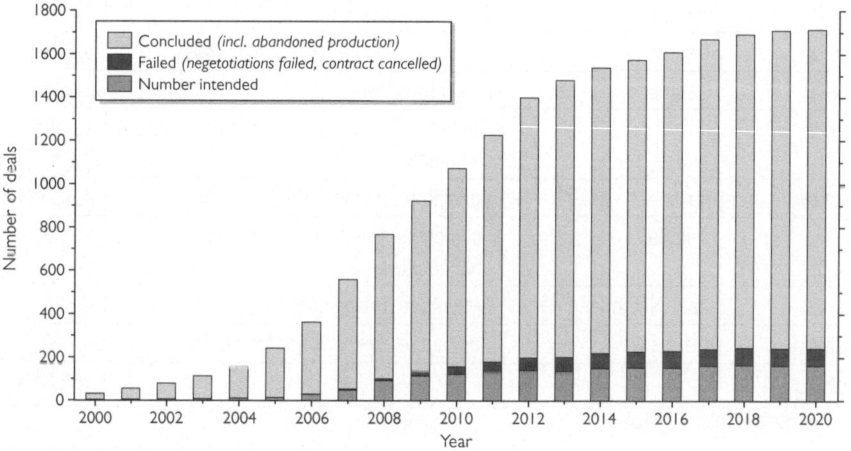

FIGURE 6.4 Transnational agricultural deals.
Source: Lay et al. (2021)

TABLE 6.2 Land Grabs by Global Region, 2016.

Region	Number of Concluded Deals	Total Size of Concluded Deals (Million Hectares)
Africa	422	10
Eastern Europe	96	5.1
Asia	305	4.9
Latin America	146	2.2
Oceania	35	26.7
Total	1,004	48.9

Source: Nolte et al. (2016)

TABLE 6.3 Land Grabs in Europe

Target Country	Investor Country	Intention	Contract Size in Hectares
Lithuania	Germany	Food crops, non-food agricultural commodities	4,650
Lithuania	Germany	Food crops, non-food agricultural commodities	3,350
Bulgaria	Bulgaria, Austria, USA	Food crops, other	21,400
Romania	Denmark	Food crops	7,536
Romania	Denmark	Food crops	1,105
Romania	Denmark	Conservation, forestry	7,261
Romania	Finland	Wood and fibre, forestry	12,000
Romania	Luxembourg	Agriculture, unspecified	105,060
Romania	France	Agriculture, unspecified, food crops	3,008
Romania	France	Agriculture, unspecified, food crops	5,500
Romania	Denmark	Food crops	3,000
Romania	Denmark	Food crops	1,200
Romania	Denmark	Agriculture, unspecified	2,000
Romania	Portugal	Biofuels, food crops	24,244
Romania	Denmark	Agriculture, unspecified, food crops	5,674
Romania	Denmark	Food crops, livestock	6,000
Romania	Italy	Food crops, livestock, non-food agricultural commodities	12,000
Romania	Italy	Agriculture, unspecified, food crops	4,850
Romania	Austria	Food crops, livestock, non-food agricultural commodities	21,000
Romania	Italy, Netherlands	Agriculture, unspecified, food crops	4,821
Romania	Germany	Agriculture, unspecified, food crops	4,700
Total			166,359

Source: European Union (2015)

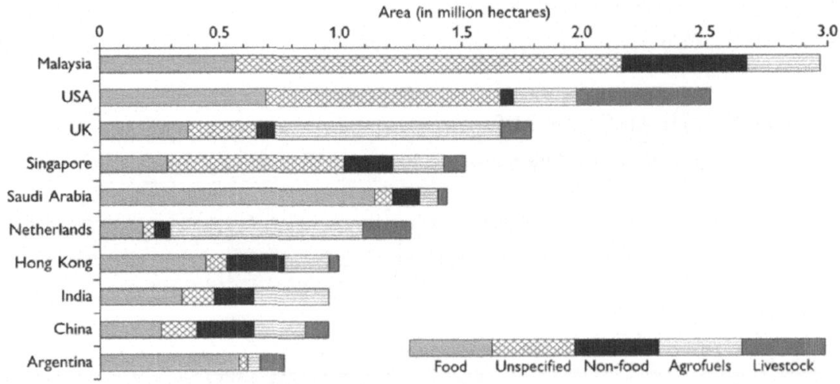

FIGURE 6.5 National investments in land grabs.
Source: Nolte et al. (2016)

It is ironic that in countries where overseas investment is significant, many farm workers have emigrated to find work in investor countries. Referring to Romania, Constantin et al. (2017: 149) comment that out of five million migrant workers in Italy, Spain, the United Kingdom, Greece, Germany and France, not one has 'bought at least one hectare of land in any of these countries, in order to set up a farm'.

6.4 Mining

Mining activities have expanded in rural areas, driven by demand from industrialising countries, rising commodity prices and new markets (Carmody, 2013). Twenty-seven times more ore and minerals were mined in 2005 than 1900, as well as 12 times more fossil fuels (Everingham, 2016). The expansion of mining has connected some previously isolated rural areas into global networks of trade (Carmody, 2013). China's Belt and Road Initiative (BRI), for example, has led to high levels of investment in some parts of Africa, leading to the export of resources, as well as the development of road, rail and other infrastructures (Mohan and Power, 2009; Power et al., 2012).

Everingham (2016) considers that there have been three main consequences for rural places:

1. New technologies are opening up new areas for exploitation. Increasingly, this is occurring in areas previously viewed as inaccessible or too challenging to mine. The clearest and most controversial example of this is hydraulic fracturing or "fracking" (Sica, 2015). This involves drilling vertically and then horizontally into shale rock before flooding it with water and sand, known as slick water, to penetrate minute fissures to release natural gas. As these technologies improve, remoter areas are being opened up for mining, such as the Athabasca tar sands in Canada (Mayer et al., 2018). The landscape and environmental aspects of these operations are controversial and have received widespread protest, both locally and internationally (Howarth et al., 2011; Vesalon and Creţan, 2015)
2. Social change. 'Mining alters, diversifies, disrupts and even displaces rural populations' (Everingham, 2016: 274). Resource booms can lead to rapid investment, employment growth and infrastructural growth, but conversely, slumps caused by changing demand and volatile prices can lead to decline and recession (Tonts et al., 2012). There have also been instances when communities have been displaced due to resource extraction (Velicu and Kaika, 2017) or the development of water resources (Heming et al., 2001). Elsewhere, close-knit communities have developed around mining activities, which continue to play an important role in maintaining social memory or national heritage (Wheeler, 2014). In some mining areas, long-distance or 'fly-in, fly-out' (FOFO) commuting is significant (McKenzie et al., 2014). Often undertaken by well-paid individuals, FOFO has been criticised for its transitory nature and lack of integration with local residents (McKenzie and Hoath, 2014).
3. Landscape and environmental impacts. Although resource extraction often accounts for a small percentage of land, it can have a profound impact on rural

landscapes and environments during and after mining operations (Wheeler, 2014). Post-mining landscapes are often referred to as 'scarred' from industrial activity, although they have also been the sites of new rounds of tourist or environmental investment. Perhaps the most famous example of this is the Eden Project in Cornwall, an international eco-visitor attraction that opened in 2001 on the site of a former clay pit (Smith, 2013). But elsewhere, a decline in primary industries in other localities has not been matched by inward investment, leading to out-migration and poverty (Rignall and Atia, 2017).

6.5 Rural Manufacturing

A surprising feature of rural economic change in the late 20th century was a growth in manufacturing, particularly in the United States and Western Europe (Healey and Ilbery, 1985; Rosenfield and Wojan, 2018; Summers, 1974). This is so much so that employment in manufacturing is relatively higher in the country than in the city in many developed nations, especially in more accessible rural places. In the UK and the USA, 9.8% and 8.9% of the rural workforce are employed in manufacturing, respectively, and in Germany, this figure is 18.5% (Rosenfield and Wojan, 2018).

The growth in manufacturing was in part driven by a decline in agricultural employment when farming became more industrialised. Given the absence of alternative employment, there was a plentiful, cheap and complicit supply of labour that could be employed in rural manufacturing. Women were a significant component of this 'green' workforce and could be employed part-time with few rights (Leach, 2016). The growth of manufacturing was also encouraged by governments through rural development programmes that aimed to diversify the rural employment base (Yarwood, 1996). These measures included the construction of industrial estates to provide premises for manufacturing, tax relief for companies (re)locating in rural places incentives for start-ups and the provision of training programmes. These inducements made rural places particularly attractive for 'branch plants' – premises that contributed to assembly lines but were dependent on headquarters elsewhere (often in urban places or overseas) (Cloke and Goodwin, 1992). While branch plants provide some employment, they are particularly vulnerable to closure in periods of recession or restructuring.

Although food processing accounts for a substantial proportion of rural manufacturing (20% in the USA) (Low et al., 2021), there is evidence that rural places are attractive to new types of industries. A rural context can support the development of creative industries that build upon and reinforce regional identities (Thomas et al., 2013) by, for example, manufacturing niche food or drink (Argent, 2018) or drawing on traditional craft activities (Yair and Schwarz, 2011). Rates of business formation can be boosted by middle-class in-migrants with the skills and experience to start new companies (Thrift, 1987; Keeble and Tyler, 1995; Westlund et al., 2014). This perhaps reflects that some rural development programmes, such as the EU's LEADER project, have stressed the importance of local capital in the development of endogenous businesses (Dax et al., 2016). Elsewhere, high-tech innovation has become synonymous with rurality through epithets such as silicon 'valley', 'bog', 'glen' and 'fen'. These rural locations offer room for expansion and space for firms to cluster but, significantly, are also close to knowledge bases and skilled workforces at prestigious universities. These advantages have led to sustained private and state investment.

Since 2000, however, there has been a decline in rural manufacturing jobs in many rural regions in OECD countries (Rosenfield and Wojan, 2018). Endogenous firms, rather than exogenous branch plants, are more likely to be sustainable (Table 6.4), especially if they manufacture niche or artisanal products associated with local rural heritage, place identity and craft tradition (Price and Hawkins, 2018).

TABLE 6.4 Causes and Solutions to Rural Manufacturing Decline (After Rosenfield and Wojan, 2018)

Trends	Challenges	Possible Solutions
Automated production	Loss of low- and mid-skilled jobs due to automation	Develop niche artisanal and craft production
Global competition	Outsourcing of assembly line tasks overseas, especially for cheap labour	Develop cultural identity of rural places to 'add value' and market locally produced goods; maintain market sustainability of local production; encourage the growth of endogenous industries rather than branch plants; develop research and design functions in rural places
Increased skill requirements	A need for a skilled workforce (often in research and development (R&D)) and marketing that are still mainly in urban places	Develop clusters of specific skills; draw on amenity value or creative place-making to rebrand places to attract and retain talent
Capacity to innovate	R&D largely located in urban places with 'branch plants' in rural areas and skilled knowledge bases (e.g., universities) in urban locations	Provide opportunities for a young, skilled workforce; develop creative clusters in rural places; create policies aimed at endogenous development
Scale and distance	Remoteness and poor broadband can hinder working and innovation	Improve digital connections; provide state support to develop technological solutions

One example is provided by Laura Ashley, who established a clothing factory and shop in Newtown, Mid-Wales, in 1961. Her eponymous company expanded to become a global brand that was synonymous with rurality and heritage (Pratt, 1992). It also became one of the biggest employers in Mid-Wales, contributing to local identity and social relations. The Laura Ashley company became public in 1985 and, following various reorganisations, was bought by a Malaysian investor in 1998 (Woods et al., 2021). It retained the factory in Newtown until it closed in 2020, when Laura Ashley plc entered administration. At the time of writing, the Laura Ashley brand has been revived by a different investment company and will open branches in Next stores. The example demonstrates that manufacturing can be significant to rural economy and society, especially when its growth is endogenous. Equally, it reveals that rural factories can be vulnerable to restructuring and global economic conditions if they are branch plants of a larger business.

Some have suggested that rural places should develop alternative local economies that make them less vulnerable to global capitalism (Gibson-Graham, 1996). McCarthy (2006b: 804) has noted that many aspects of 'alternative' economies, such as 'more face-to-face interactions, less physical and social distance between production and consumption' are particularly suited to rural settings. One example of this is in Totnes, Devon, where activists have attempted to develop a 'circular economy' based on recycling resources, supporting local produce and the use of local currency (Hopkins, 2008; Bailey et al., 2010). The aim is to improve social, economic and environmental resilience by transitioning away from fossil fuels and a reliance on the global economy.

6.6 Conclusions

The previous two chapters have examined some of the broad economic and structural changes that have impacted rural areas. It is important to note that these changes are uneven and affect different places in different ways. Some places, for example, are still dominated by productive agriculture, others are driven by amenity, tourism and leisure, and others may be 'left behind' when formerly dominant industries collapse. Consequently, economic change can be explained by a number of key drivers (Table 6.5) that vary in significance between different rural places (Shucksmith and Brown, 2018).

TABLE 6.5 Different Explanations of Rural Economic Change (After Shucksmith and Brown, 2018)

Key Explanations	Summary of Change
Agri-centric change	Rural areas are essentially agricultural, with farming privileged as the key economic driver of change and the main form of social organisation.
Urban-rural access	Access to urban places determines change, especially commuting distance or 'travel to work areas'. More accessible urban areas experience wider and more profound changes.
Competitive economy	Rural places are affected by global competition and capital. Thus, compliant, low-skilled and cheap labour may appeal to global investors; areas of skilled populations may lead to entrepreneurial development and investment.
Places left behind	Places, particularly those previously dependent on extractive or primary industries, are in decline, with degraded landscapes and poor communities.
Amenity-based economic narrative	Places with attractive natural amenities appeal to new forms of travel, leisure and tourism, as well as gentrifiers and second home owners.
A narrative of society-nature relationships	Rural society and the environment are affected by changing natural environments, particularly in light of climate change and sea-level rise. Some benefit, for example, through longer growing seasons, while others are threatened by environmental degradation.

Change is not a one-way process and, as Halfacree (2007) reminds us, can be coherent, contradictory or chaotic. In some localities, change is accepted, while in others, it is challenged or contested. Change is not, therefore, predetermined but, instead, reflects an interaction between global processes and local social conditions. In order to better understand rural change, it is therefore important to consider how change is imagined, contested and played out by people in their daily lives. With this in mind, the following chapter considers some of the ways that the countryside has been socially constructed and how different ideas of rurality influence change.

CHAPTER

7 Changing Representations of Rurality

7.1 Introduction

Despite the momentous transformations described in the previous chapters, the country is widely seen as unchanging and idyllic (Box 7.1). Raymond Williams (1973: 297) comments that 'the common image of the country is now an image of the past, and the common image of the city an image of the future'. The country has been associated with a 'rural idyll' or 'ideal' (Bunce, 1994; Gold and Revill, 2004; Mingay, 1989) that John Short (1991) characterised as:

- nostalgic;
- part of national identity;
- traditional;
- problem-free;
- close-knit/friendly;
- a place of play;
- simple;
- natural.

The rural idyll has been an influential idea in many countries and has had an important bearing on the ways that the countryside has changed. This chapter starts by considering the significance of the rural ideal before going on to consider alternative ways in which rurality is socially constructed (Halfacree, 1993).

Box 7.1 An Example of the Rural Idyll

The continuing power of a nostalgic rural idyll is evident in a recent advertising campaign by a UK train company, Great Western Railway, that used Enid Blyton's Famous Five stories to promote train travel.

DOI: 10.4324/9780429448966-9

FIGURE 7.1 GWR and the rural idyll.
Source: Great Western Railway

Blyton's books, written between 1942 and 1963, tell the adventures of the eponymous four children and their dog in the English countryside and Dorset in particular (Jones, 1997). Theirs is a world of adventure and exploration through hearty outdoor activities, such as bike rides, walks, picnics and camping, punctuated only by crimes to solve and mysteries to unravel. The books themselves were written in a nostalgic fashion, often ignoring modern technologies or events, and, despite criticism and parody for their staid portrayal of class and gender, remained popular for their celebration of freedom and fun in the countryside.

The Famous Five, including Timmy the dog and a picnic, are recognisable in the advert, which is supported by the strapline of 'Five enjoy a spring break'. The use of these characters persuades the viewer to see the countryside in a nostalgic way. This wistfulness, in turn, encourages people to rediscover idyllic, fun and simple days out in the country. The whole point of the advert is, of course, to boost travel by rail. Here, the train, the latest Hitachi Class 800, is the only evidence of 21st-century life but is positioned in the background of the picture, almost unnoticed. The suggestion is that train travel is part of the nostalgic experience and is necessary to enjoy a return to an imagined rural past. The train, despite or because of its sleek, modern and Japanese design, blends seamlessly into the picture, confirming that it is part of the fabric of rural railways that, in turn, have been associated with national identity by, for example, the former poet laureate John Betjeman. Although the realities of rural life and British train travel are very different, the poster confirms that the countryside is unchanging and untrammelled.

7.2 The Rural Idyll: A Nostalgic History

The idea that the country is idyllic is far from new. Writers in ancient Greece yearned for a lost golden age characterised by abundant crops, docile animals, fertile soils and lives of leisure (Short, 1991). Theocritus' bucolic poetry celebrated simple, peasant farming practices in the region of Arcadia, giving rise to the term 'Arcadian' to describe pastoral landscapes and lives. Yet Arcadia itself was a harsh mountainous environment where agricultural toil, rather than idyllic existence, was necessary for survival (Short, 2006). Even in the classical world, Arcadia was a 'dreamscape . . . which stands in dramatic contrast to the experience of everyday life' (Swaffield and Fairweather, 1998: 113).

In his seminal book, *The Country and the City*, Raymond Williams (1973) demonstrates that writers throughout history have looked back at a rural golden age that appeared to exist just prior to their time of writing. Despite being tantalisingly out of reach, the rural idyll has remained an incredibly powerful and significant idea (Hall, 2020).

Different versions of the rural idyll exist in other countries. In some cases, Arcadian ideals were transplanted by and for European settlers in American and Australasia (Swaffield and Fairweather, 1998). In 19th-century New Zealand, for example, efforts to colonise the countryside around Christchurch rested on efforts to re-establish a rural society that had been lost in industrial Britain. Incentives were offered to re-establish an agrarian social order based on ownership of land and social hierarchy (Swaffield and Fairweather, 1998). In Australia, it was envisaged that:

> *for £150, aspiring colonists would get a 20 ha rural allotment and a half acre urban section. They would form the gentry, while farm servants, shepherds, domestic servants, country mechanics and artisans would receive assisted passages to provide labour for the new colony.*
>
> *(Swaffield and Fairweather, 1998: 115)*

In both cases, Arcadian ideals were to be enjoyed by rural elites in ways that reflected their power.

Today, the rural ideal in New Zealand is widely expressed through 'lifestyle blocks' or small holdings that afford urban migrants opportunities to farm or work on the land, albeit on a part-time basis. The Arcadian convention therefore continues to maintain its potency, even half a world from its place of origin, and remains a significant factor in the development of rural land use in the urban fringe

(Swaffield and Fairweather, 1998: 124). Likewise, in the USA, scarce labour and vast amounts of land (taken from Native Americans) led to a rural society dominated by family farms that valued independence, a strong work ethic and a frontier spirit (Bunce, 1994; Short, 1991). A nostalgia for 'a less complicated, pre-industrial understanding of home and "pure" nature' continues to drive migration to some parts of the United States (Nelson and Nelson, 2011: 444).

7.3 The Importance of Representations

The critic AA Gill (2005) once remarked, 'Britain is a country that is best seen by drawing the curtains, opening a book and never leaving the room'. In the Global North, the majority of people live in urban areas and so experience an 'armchair countryside' (Bunce, 1994) through various forms of media (Short, 2006), from high art to Facebook. These representations develop an idealised, general view of the country: 'the idea of a type of place rather than a specific place attachment' (Harrington, 2018: 250). The imagined country has been traced in various media, including TV programmes (Box 7.3) (Horton, 2008; Phillips et al., 2001), music (Yarwood and Charlton, 2009), books (Chueh and Lu, 2018) and films (Fish, 2007), confirming that, for many people, rurality is experienced more in the mind more than in reality (Mormont, 1987; Crang, 2015).

These representations of the countryside are significant. They are not only a way of seeing the countryside but also reflect an ordering of people, landscapes, cultural norms and social practices that assert normative ideals and power relationships. Idyllic representations of the countryside have had four important consequences:

1. *The idyll has contributed to key rural processes.* A desire to live in idyllic rural places has contributed to urban-to-rural migration in many Western countries (Barcus and Halfacree, 2017). Work suggests that migrants strive to buy into and maintain an idealised vision of rurality, often joining local political organisations or conservation groups to preserve idyllic views or lifestyles (Phillips and Smith, 2018b; Phillips, 2014; Woods, 2006). Consequently, a desire to maintain a picturesque landscape often takes priority over the need to build affordable housing (Gallent et al., 2015).

2. Numerous studies have demonstrated that *idealised visions of rural life have papered over social problems* and prevented their solution (Vepsäläinen and Pitkänen, 2010; Cloke et al., 1995a; McLaughlin, 1986a; Shucksmith, 2018; Matthews et al., 2000; Yarwood, 2005a). It is often assumed that because an area looks idyllic, it is also problem free (Cloke et al., 1995a; Milbourne, 2014). As a consequence, policy interventions are more likely to be aimed at urban areas where social disadvantage is more visible (Cloke, 1999).

3. The idyll *has determined who fits into rural places* and the kind of behaviours expected in them (Cloke and Little, 1997). At its most extreme, there is a historical example of a village being razed and built elsewhere in response to school children stealing fruit from the landowner (Short, 1991)! More subtly, though, hegemonic visions of rurality contribute to an ordering of the country that implicitly includes some people, activities and landscapes while, at the same

time, glosses over things that transgress or disrupt this image (Box 7.2). Travellers, for example, have often been regarded with suspicion (Holloway, 2007), young people 'hanging around' can be seen as a threat (Valentine et al., 2008; Yarwood and Gardner, 2000), and asylum seekers have been viewed as criminals (Hubbard, 2005). Women are often expected to behave in ways that conform to a gendered stereotype of rural life by, for example, demonstrating support for (male) farmers or being proficient in domestic skills (Little, 2002a; Little and Austin, 1996) (see Chapter 17).

4. Finally, the image of rural places has been *crucial to rural restructuring*, especially in the development of tourism (Barrett et al., 2001). Place marketing has become more significant as rural places seek to identify and market what makes them different to other places. Particular niches, such as food (Bessiere, 1998; Sims, 2010), music (Gibson and Davidson, 2004), heritage (Storey, 2016), adventure (Cloke and Perkins, 1998), agriculture (Harrington, 2018), arts (Mahon and Hyyryläinen, 2019) or wildlife (Cloke and Perkins, 2005) are important in the development of place-branding. Other places appeal to associations with television or film, such as Heriot Country in Yorkshire and Lord of the Rings Country in New Zealand (Phillips et al., 2001). Kurow in New Zealand welcomes tourists to 'Richie Macaw Country', named after the iconic All Black captain who was born in the region (Roy, 2015)!

Box 7.2 Rural Rebranding

Places are increasingly being viewed as 'brands' that can be used to attract investment (Medway and Warnaby, 2014). Although place-branding was originally used in the post-industrial transition of urban places (Hall and Hubbard, 1996; Kearns and Philo, 1993), rural places, no matter how small (Figure 7.2), have also sought to brand themselves in distinctive ways in order to attract tourists, visitors or investment (Storey, 2016). Urry (1990: 125) demonstrated that 'all sorts of places (indeed almost everywhere) have come to construct themselves as objects of the tourist gaze – in other words, not as centres of production or symbols of power but as sites of pleasure'. Branding involves analysis of issues, a strategy aimed at developing and promoting a place-brand and, crucially, a participatory phase in which activities in the place align with its new identity (Ashworth and Kavaratzis, 2015).

One famous example of this approach can be seen in Tamworth in New South Wales, Australia. The country town has become a significant centre for country music (Gibson and Davidson, 2004) through a large recording industry and a huge annual festival. As a consequence, Tamworth has branded itself as the Country Music Capital of Australia. In doing so, Tamworth appropriated ideas of the country as 'credible', 'decent', 'honest' and 'consistent' and bound these into existing ideas of Australian rural identity (Gibson and Davidson, 2004), mate-ship and

FIGURE 7.2 Place marketing. Widecombe in the Moor on Dartmoor draws on a famous folk song 'Widecombe Fair', in which Uncle Tom Cobley rides an old grey mare with a host of characters.

stoicism (Cunneen, 2001). Promotional material largely drew upon and reinforced gendered ideals of rural work or sports, reinforcing masculine notions of 'taming nature' (Gibson and Davidson, 2004: 396). By contrast, women, when they are pictured, are either passive or in domestic roles. Indigenous Australians are absent, despite there being a significant community in or around Tamworth. Local residents have also objected to the association of Tamworth with a 'hick' image: 'it's very hard to shove off the stereotypical image of country people – that you've got to be a redneck, like country music, and be able to linedance' (resident quoted in Gibson and Davidson, 2004: 400). Despite the commercial success of promoting Tamworth as the Country Capital, this image is problematic as it hides as much as it reveals.

In the English Midlands, ideas of heritage have been used to 'name' and 'theme' long-distance walking routes to link different places together in order to attract visitors on foot (Storey, 2016). Examples include the Mortimer Trail, named after an influential medieval family, and the Simon Evans Way, named in honour of a local poet and postal worker. Cornwall has become a popular destination for German visitors who are keen to see where the British novelist Rosamunde Pilcher set her novels. Although she is little known in the UK, it is estimated that her work attracts over a quarter of a million German tourists a year to Cornwall (Jakat, 2013).

7.4 The Anti-Idyll

Although versions of the rural ideal often dominate rural places, they rarely go unchallenged (Bell, 2006; Short, 1991). Tim Hall (2020) points out that there has been significant growth in anti-idyllic media since the 1960s and, drawing on a survey of anti-idyllic British media, he identifies the following themes:

1. Landscapes that are 'grim, ugly, bleak or isolated' (Hall, 2020: 8) provide backdrops to crime stories (Brunsdale, 2016), as Nordic noir television demonstrates (Bergman, 2014). If idyllic landscapes are portrayed, they are used as a counterpoint to activities that seem more shocking because they occur in such bucolic conditions. *Broadchurch*, for example, is set in a Dorset coastal town and draws on shots of rolling landscapes that stand in contrast to a brutal murder.
2. Rural communities and rural life are portrayed as 'small, closed, insular and unchanging, characterised by close, hierarchical familial and social bonds' (Hall, 2020: 11). Rather than virtues of rural life, these traits contribute to boredom, lack of opportunity and a feeling of being trapped, especially for young people. The television programme *This Country* centres on two cousins with little prospects living on a council estate in the Cotswolds (Box 7.3).
3. Rural labour and the rural economy are viewed as 'repetitive, squalid and soul-destroying' (Hall, 2020: 13). Jonathan Coe's novel *What a Carve Up!* uses battery farms to portray the countryside as 'a site of industrialised cruelty and death'.
4. The lived cultures of the non-idyllic countryside are portrayed in some media. This includes 'modern slavery and exploitative labour practices; fly tipping (the illegal dumping of waste); illegal blood sports; illegal music events; and "dogging" (meetings of groups of strangers in secluded locations for casual sex)' (Hall, 2020: 14). The world's longest soap opera, *The Archers*, has delved into problems such as alcoholism, domestic violence, homelessness, dementia and farm accidents.

Box 7.3 A Televisual Tale of Two Villages

The Detectorists and *This Country* are two successful but different sitcoms that draw upon English rural life.

The Detectorists follows the life of Andy (Mackenzie Crook) and Lance (Toby Jones), who enjoy the hobby of searching for historical artefacts with metal detectors. It is an acutely well-observed tale of male friendship and hobby cultures, which is set in the fictional village of Danebury in Essex. It has a deliberately slow pace that includes lingering shots of wildlife, plants, insects and sunsets that celebrate and draw upon Danebury's rural setting. Andy and Lance both work jobs below their abilities and intelligence, although this is partly by choice as it allows time for their hobby. The show has attracted attention from geographers, partly because they are fans, but also because it 'spoke to the rural environment at a time of ecological crisis', celebrated stillness and revealed contested ideas of Englishness in a rural context (Innes and Norcup, 2020: 15). The rural landscape is more than a backdrop for the show but is 'walked, surveyed, sensed, gazed upon, read, and dug. Landscape is where the characters seek solitude, find companionship, and navigate the sometimes dramatic intrusions from the "the rude world"' (p. 17). As such, rurality is portrayed as something idyllic, to be celebrated, savoured and saved from urban intrusion (Harries, 2020). Throughout the show, modernity is connected to the past by the hobby of detecting. Change is accepted by the characters, but, at the same time, the programme reveals a mournfulness and sense of ghostliness associated with the country. This is also captured in the programme's excellent theme tune, a folk song by Johnny Flynn, that recalls 'I'm with the ghosts of the men who can never sing again'. Mackenzie Crook, who wrote and directed *The Detectorists*, commented on finding a hawking whistle:

> *[I] held it to my lips and blew. The note that issued from the whistle was a ghost, a sound unheard for centuries . . . and it wasn't a faint, feeble ghost either: it was an urgent piercing shrill that echoed across the field and back through time.*
>
> *(Crook, 2020: 11–12)*

The Detectorists portrays rurality in an idyllic yet subtle way that is reflected in its slow pace and portrayal of landscape and the past. At the same time, it recognises that these are an escape from the daily life of work, progress and change. Yet in doing so, it does not seek to satirise change, reveal a darker countryside or wallow in the past; it simply celebrates what can be found by engaging with landscapes with particular knowledge and in particular ways (Innes, 2020).

Likewise, *This Country* also works by contrasting urban lifestyles with rural realities, albeit in very different ways! The series follows Kerry and Lee 'Kurtan' Mucklow (Daisy May Cooper and Charlie Cooper), two disenfranchised young people who live their lives on a social housing estate in the otherwise picturesque Cotswolds. The programme plays on a sense of tragi-comedy (Hall, 2020) that derives from the delusional ways in which they view their home village (Brooks, 2018). Kerry's talk of gangs and organised crime, usually associated with inner cities, derives its comedic value by name-checking places more readily thought of as twee:

> *There are people from my past who would love to see me slain. I got enemies everywhere. I got enemies in South Cerney. I've got enemies in North Cerney. I've got enemies in Cerney Wick. I got enemies in Bourton on the Water. There's a tea room there and under the counter they got a panic button, if I take one step inside they can press that and police will be there in three minutes.*
>
> *(Quoted in Hall, 2020: 12)*

A sense of bathos is achieved when these imagined urban lifestyles, together with excitements of seeing minor celebrities in their local shop, is contrasted with the boredom of the protagonists' jobless and mundane daily lives, something that reflected the Coopers' own lives and frustration at being trapped in a place with little opportunity (Brooks, 2018).

Neither *This Country* nor *The Detectorists* satirise rural life, yet both play upon, unpack and challenge widely held assumptions about the countryside. In doing so, they present a nuanced take on the rural idyll and the gaps between its imagination and reality.

7.5 Conclusions

This chapter has briefly illustrated some of the dominant ways in which the countryside has been portrayed in various media. Although the idea of the rural myth continues to be significant, it has been challenged by other portrayals of rural places. Although anti-idyllic ideas are gaining purchase, they do not themselves offer a more 'realistic' version of rural life. Rather than considering whether a portrayal of the countryside is true or false, it is more important to consider whose truth it is (Short, 1991). Consequently, greater attention has been paid to listening to different voices or versions of rural life (Chapter 14) and not privileging some accounts over others.

CHAPTER

8

Changing Rural Lives

8.1 Introduction

The previous chapters used Halfacree's (2007) model of rurality to examine how the country has changed since the Second World War. They traced how broad agricultural and economic changes have altered rural localities and considered how the countryside is represented in different ways. The final part of Halfacree's (2007) triumvirate considers the everyday lives of people living in the country. It is tempting to consider social change as an outcome of rural restructuring and the cultural reimagination of the countryside. While recognising that these processes have had a profound impact on rural life, Halfacree's model also stresses that people and communities themselves shape rural change. In other words, people are not passive receivers of change but are active agents in the way it occurs. To explore these issues, this chapter examines some of the key social changes occurring in rural areas. It starts by considering migration to rural areas, as this is both a cause and effect of social change.

8.2 Demographic Change

Urbanisation is the dominant trend in many parts of the world. In 2009, the number of people living in urban areas (3.42 billion) surpassed those in rural areas (3.41 billion) for the first time. In 2020, 55% of the world's population lived in cities, and this is set to rise to 68% by 2050. Ninety per cent of this growth will occur in Asia and Africa and, specifically, India, China and Nigeria (United Nations, 2020a). In Western countries, this was also the case until about the 1970s. As agricultural employment fell (Clout, 1980), younger people moved to cities in search of work, housing and better opportunities, leaving an ageing population behind (Weekley, 1988; Newby, 1979b). Yet, in the middle of that decade, Brian Berry (1976) identified that the rural population of America was starting to grow and there was a corresponding decline in the urban population. Berry declared that:

DOI: 10.4324/9780429448966-10

a turning point has been reached in the American urban experience. Counterurbanisation has replaced urbanisation as the dominant force shaping the nation's settlement patterns.
(Berry, 1976: 17)

Since then, counterurbanisation has been the prevailing trend in most Western countries (Figure 8.1) (Boyle and Halfacree, 1998; Champion, 1989; Mitchell, 2004; Bosworth and Bat Finke, 2020; Rowe and Patias, 2020; Barcus and Halfacree, 2017). It is characterised by migration from urban to rural areas, often by middle-aged, middle-class people, leading to a net increase in rural population and a net decline in urban populations.

There has been some debate about the extent to which counterurbanisation represents a 'clean break' from urban living or whether it is a form of suburbanisation that has been enabled by improved transport or, more recently, e-working (Champion, 2001; Mitchell, 2004). The most accessible rural areas have indeed experienced the highest increases in population, but the trend also appears to have affected remoter areas too, suggesting that the relative importance of 'suburbanisation' and 'clean break' migration depends on proximity to urban places.

In all locations, one of the most significant drivers of urban-to-rural migration is the desire to adopt a rural lifestyle. This is associated with nature, safety, slower lifestyles and close-knit communities (Benson and O'Reilly, 2009; Walmsley et al., 1998; Halfacree, 1994). More specifically, some migrants move to provide a better environment for children and family (Bushin, 2009), adopt alternative lifestyles (Halfacree, 2006, 2009), establish niche businesses (Argent et al., 2013), return to childhood homes (Laoire, 2007), be closer to family (Gkartzios, 2013) or live in places familiar to migrants through visits or holidays (Drysdale, 1991). There is evidence that some places have attracted particular migrants for specific reasons: for example, Hebden Bridge in West Yorkshire has drawn in gay people (Blackstock et al., 2006) and Totnes in Devon has appealed to those seeking environmentally

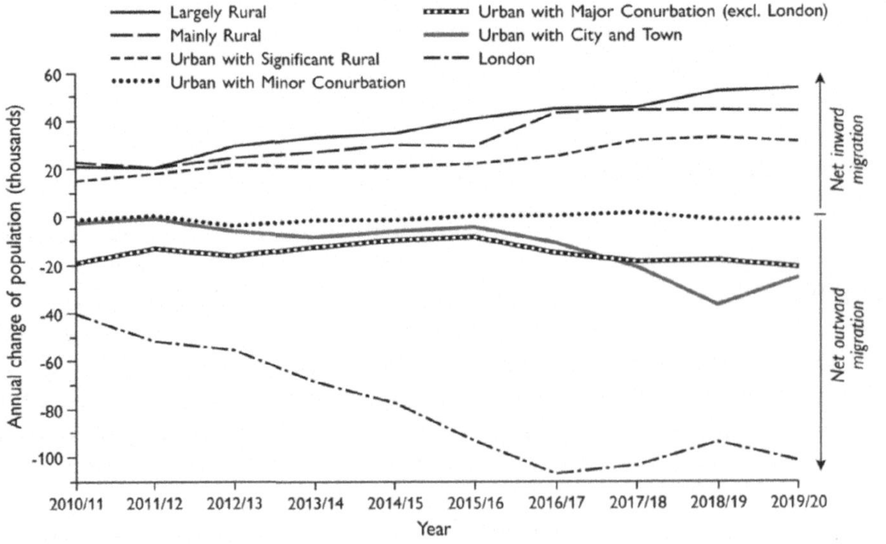

FIGURE 8.1 Internal migration in England.
Source: Department for Environment, Food and Rural Affairs (2022)

friendly lifestyles (Hopkins, 2008). In Australia, coastal locations are significant to lifestyle migrants (Curry et al., 2001; Walmsley et al., 1998); wilderness is important in Canada (Chipeniuk, 2004); and feelings of security attract movers to Scandinavia (Persson, 2019). The desire to move to the country is supported by popular culture with television programmes such as *Escape to the Country* or *Location, Location, Location* following movers to idyllic places (Halfacree, 2008).

There is also evidence that some migrants seek rural idylls in other countries. Sweden has proved attractive to Dutch and German migrants (Eimermann, 2013; Persson, 2019), France to the British (Benson, 2013; Ferbrache and Yarwood, 2015) and South America to US citizens (Rainer, 2019). One international migrant describes the process:

> *We used to go on holiday to France, and then some friends recommended that we visit Sweden; so we decided to give it a try. And when we got here, we thought 'wow, this is actually what we have been looking for in France' – the same abundance of wilderness, but with the right of public access. In France, everything is gated. And here, you have access and can enjoy the forests and the lakes. We were sold instantly! So tranquil and beautiful.*
>
> *(Interview 8, 2011 quoted in Eimermann, 2013)*

It is not only important to consider *why* people want to move to the country but also who is *able* to move (Stockdale, 2014). Wealth is the main determinant. Given a lack of affordable housing (Chapter 11), movers to rural places need the capital to buy a house in an already competitive market. If their move has originated in an expensive urban area (such as London), these migrants may be able to outbid local people for housing, further inflating house prices. Consequently, middle-class home-owners are more likely to move to rural places (Stockdale et al., 2000). Professional or managerial jobs also offer more opportunities to work flexibly from home or commute longer distances (Pahl, 1965a; Cloke and Thrift, 1987; Lowe and Ward, 2009). Indeed, for some, a longer commute from a home in the country may be cheaper than urban living, even with the additional travel costs.

The decentralisation of businesses, rural manufacturing growth, establishment of branch plants, improvements in broadband provision and increase in service sector employment have also created opportunities for some people to move the country (Cloke and Goodwin, 1992; Green, 2016; Coombes et al., 1989; Thrift, 1987). Counterurbanisation is therefore a characteristic of the post-industrial economy (Bosworth and Bat Finke, 2020), and, by contrast, rural-to-urban migration, or urbanisation, is dominant in countries that are industrialising, such as China. It also has been suggested that rural migrants are important in the formation of new businesses and are therefore important contributors to post-industrial economic growth (Stockdale, 2006). These may be associated with the development of new products or services that use ideas on locality and rurality to develop niche markets (Bosworth, 2010; Hoey, 2005; Bell and Jayne, 2010; Argent et al., 2013).

Stage in the life cycle is also a significant driver of counterurbanisation. Some families migrate to provide a better environment for young children (Bushin, 2009) and people who have retired, or are planning to retire, may seize the opportunity to move to the country or coastal locations (Brown and Glasgow, 2008; Weekley, 1988). Again, wealth and class determine who is able to make these moves.

By contrast, working-class people are less likely to move to rural places (Halfacree, 2008). One exception is Princetown in West Devon. The town originated at the start of the 19th century to provide housing for prison officers working at Dartmoor prison. Until 2001, officers were required to live in the town, but, following the relaxation of this regulation, many left Princetown, and their state-owned accommodation was instead used to house families from nearby urban areas. Consequently, Princetown's age profile is much lower than other settlements in Dartmoor; it has some of the lowest house prices in Dartmoor and relatively few second homes (Dartmoor National Park 2019). Limited local employment and poor transport links put these movers at higher risk of social disadvantage. Their migratory experience of moving to the country is therefore in stark contrast to middle-class migrants. This highlights a need to understand diverse forms of migration from marginalised groups. Others include seasonal migrant workers (Halfacree and Rivera, 2012) or those who move for alternative or countercultural idylls (Halfacree, 2009).

Lack of opportunity and isolation also mean that some people are more likely to move out of the countryside. In particular, there continues to be a net movement of younger people from rural places in search of employment, education and excitement (Haugen and Villa, 2006; Stockdale, 2004). In England in 2017, for example, the net figure of 88,400 migrants to rural areas hides a loss of 41,400 people aged 17–20 (Figure 8.2), largely to take up places in higher education institutions in urban areas. In the USA, it has been observed that 46% of remote rural counties are depopulating (compared to 24% of the adjacent nonmetropolitan counties and 6% of metropolitan counties) (Johnson and Lichter, 2019). As Figure 8.3 reveals, population decline reflects employment losses in counties reliant on agriculture, mining, forestry and manufacturing (Johnson and Lichter, 2019). By contrast, areas of high amenity, such as the Great Lakes, New England and the Great Smoky Mountains, continue to attract migrants.

As a result of these migratory trends, and a consequent decline in fertility, rural populations are growing older (Glasgow and Brown, 2012; Burholt and Dobbs,

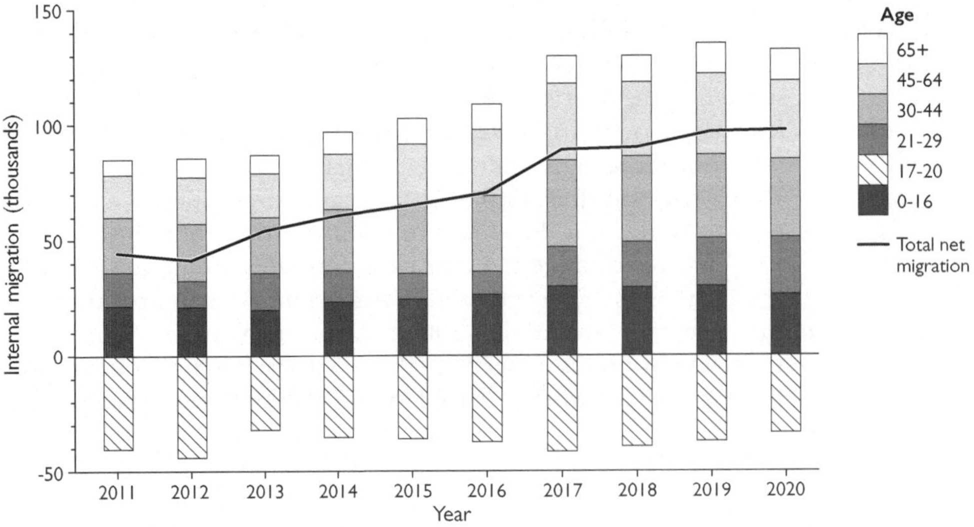

FIGURE 8.2 Internal migration by age in England.
Source: Department for Environment, Food and Rural Affairs (2022)

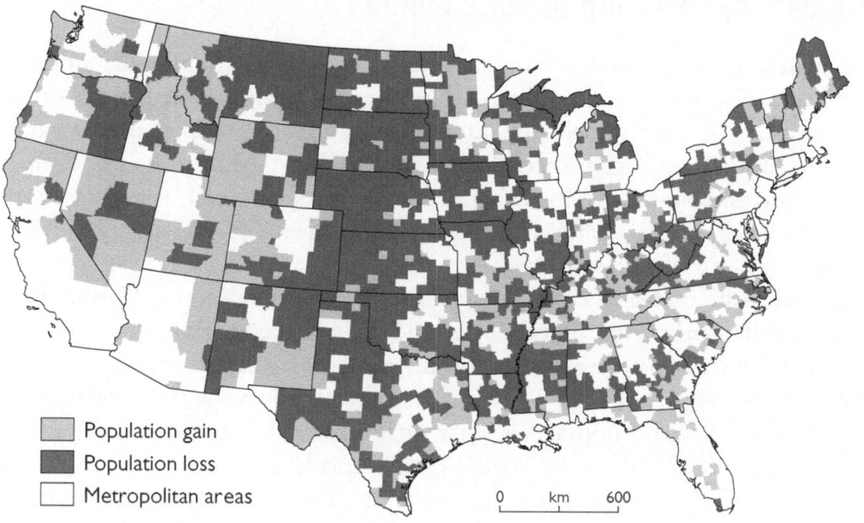

FIGURE 8.3 Population change in rural USA.
Source: Johnson and Lichter (2019)

2012) (Table 8.2). In England, 23% of residents in rural districts were aged 65 years or over in 2007, and it is estimated that this will increase by 62% by 2029. Once rather cruelly referred to as 'geriatrification' (Weekley, 1988), ageing brings considerable challenges in rural places. Loneliness, health care and isolation from services are particularly pressing issues (Milbourne and Doheny, 2012). Dementia, for example, is set to increase in rural places, challenging the provision of health care (Kelly and Yarwood, 2018).

Furthermore, ageing has led to a significant increase in the number of rural *households* (Lewis, 1998). As people live longer, there has been an increase in single- or dual-occupancy households that have impacted the availability of housing. For example, a widow may continue living in a four-bedroom house because it is her home and has an emotional significance to her. But this means that a suitably sized house is not available for a young family to buy. If this family moves away to find a house, a situation may occur whereby the number of households remains constant but the population declines. Increases in divorce and delays in the age of marriage have also contributed to the growth of single-person households (Barcus and Halfacree, 2017).

An emphasis on long-distance migration has also hidden the importance of local movement, which is very significant in most places (Milbourne, 2007). Indeed, counterurbanisation has sometimes been seen as a series of shorter moves up, down and across the settlement hierarchy (Fielding, 1989). For example, as migrants become older and less able to drive, they may seek to move from isolated rural idylls, which initially attracted them as 'young retired' migrants, to larger villages or towns with better access to health care and services.

Demographic transition is, therefore, complex and highly geographical. Population change are a result of economic restructuring outlined in the previous chapters, as well as how the countryside is variously perceived as an idyll or backwater. Migration therefore reflects and affects the changing social structure of many communities.

8.3 Gentrification, Class and Cultural Conflict

Counterurbanisation is both a cause and effect of rural change. As noted in the previous section, it reflects changing economic structures and new ways of imagining the country. At the same time, the countryside is being shaped to meet the expectations and spending power of new populations. Although it is tempting to see communities divided by 'locals' and 'newcomers' (indeed, many lay conversations frame change in this way), it is important to view these changes in relation to wider changes in rural society.

In the past, ownership of land and agrarian practices relied on a social hierarchy in which people worked for a landowner and, in return, received paternalistic benefits, such as housing and, to some extent, forms of welfare through charity and benevolence (Newby et al., 1978). It was also reflected in the form of rural politics that was associated with maintaining this agricultural status quo and the interests of large landowners (Gallent et al., 2008a; Curry and Owen, 2009; Winter, 1996). In the post-war period, these hegemonies were challenged as farming changed. New economic activities brought in new divisions of labour, with less reliance on agriculture and its social structures. Perhaps more significantly, counterurbanisation by the service classes resulted in new patterns of property ownership and new forms of politics to protect their investments and interests (Woods, 2006).

The term service class refers to those who are 'neither exploited workers nor owners of capital . . . *but service capital* by providing specialist skills' (Woods, 2004: 85). These groups are well paid but, significantly, have considerable autonomy that enables flexible or home working. As noted in the previous section, the service classes have the means and inclination to undertake counterurbanisation and purchase rural properties (Cloke and Goodwin, 1992), which can reflect a form of *gentrification* (Box 8.1). The service classes also have the power to change rural places to meet their expectations by, for example, lobbying planning meetings to prevent intrusive developments (Cloke et al., 1995b).

Box 8.1 Rural Gentrification

Gentrification has been described as 'a change in the social composition of an area with members of a middle-class group replacing working-class residents' (Phillips, 1993: 124) and has been more usually applied to urban areas (Phillips and Smith, 2018a). It has been associated with three processes: the social transformation of an area involving the middle classes, the displacement of former residents and the refurbishment of properties (Stockdale, 2010). Two main theories have been advanced for gentrification:

1. The rent-gap theory based on uneven development (Smith, 1979). Capital is invested in some areas (such as the suburbs) and withdrawn in others (such as inner-city areas), where property prices subsequently fall. Falling property prices then make these places attractive to investors, especially if they are part of a regeneration programme. In a rural context, economic decline, agricultural restructuring and

out-migration contributed to a fall in property values in some areas (Newby 1979b). For some, these properties offered good investment opportunities. In one study in the Gower, nearly a third of properties had undergone improvements that included adding new rooms, knocking down internal walls or renovating outbuildings (Phillips, 1993), with a view to improving property value and resale.

2. Demand or consumption-led investment (Ley, 1987) in which new middle classes seek housing in places with character, cultural diversity and shorter commutes as part of a lifestyle choice. Some people move to the country to achieve particular rural lifestyles and, once there, alter properties and places to match these expectations (Phillips, 2002).

Stockdale (2010: 39) cautions that it is 'important not to confuse gentrification with other migration processes'. Questions have also been raised about the extent to which gentrification represents a displacement of existing residents. In a study of Berkshire, Phillips (2002) reveals that new service-class families tended to move into newly constructed homes rather than existing properties. Hence, they added to, rather than dislodged, existing populations from the area. Cloke and Thrift (1987) also consider that rural gentrification reflects change and contest between different middle-class factions rather than between middle- and working-class populations.

Gentrification also reflects the importance of different forms of capital (Phillips and Smith, 2018a, 2018b) (Figure 8.4). High investment of eco-

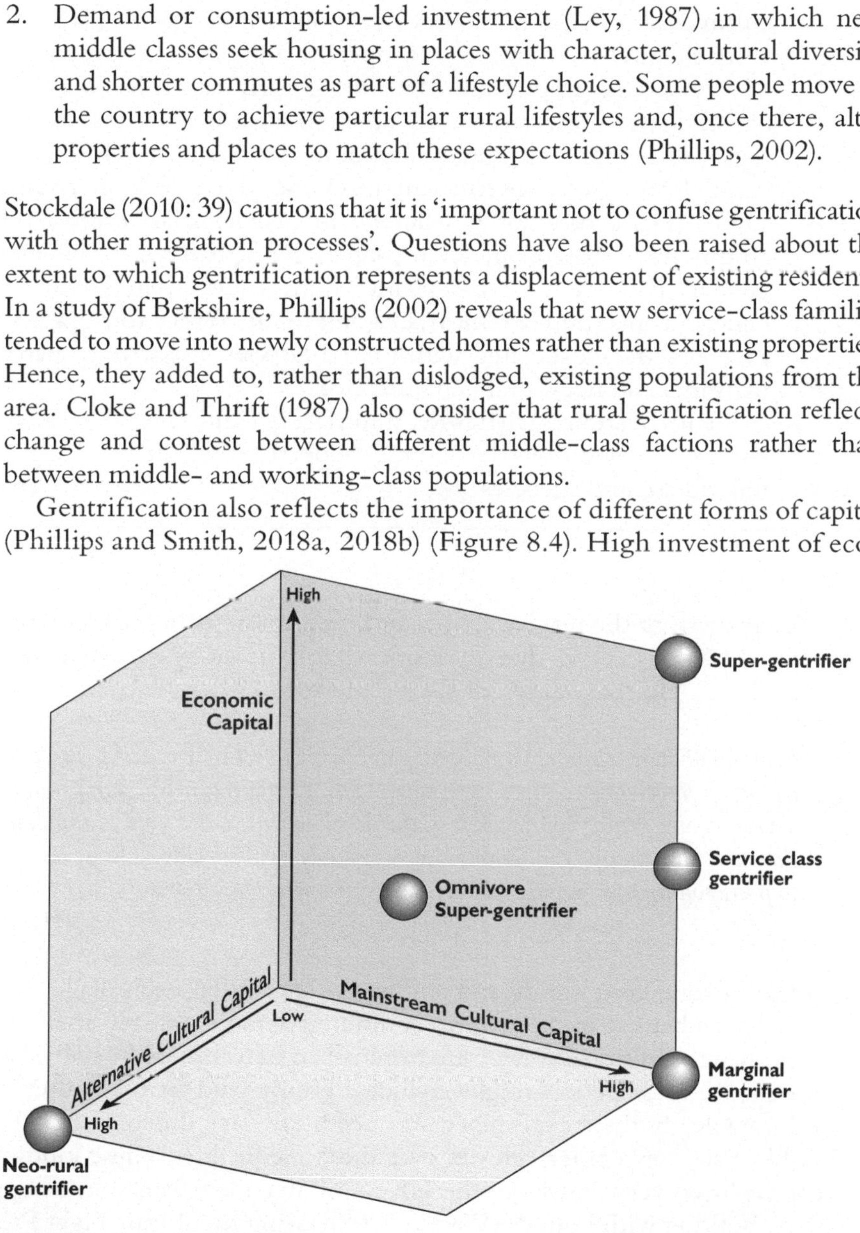

FIGURE 8.4 Theories of gentrification.
Source: Phillips and Smith (2018a)

nomic capital is reflected in super-gentrification, in which gentrification extends beyond residential properties to encompass shopping, restaurants and cultural facilities to create even more exclusive and expensive enclaves. Examples include gated communities or resort development, especially overseas (Rainer, 2019). By contrast, marginal gentrifiers are less-privileged members of the middle classes who may have limited amounts of economic capital but relatively high levels of educational or cultural capital. As such, they draw on 'sweat equity' (i.e., carrying out work themselves) to improve properties and pave the way for other forms of service class migration. Marginal gentrification has been undertaken by students, creative people or single parents attracted to what are regarded as plural or tolerant communities (Rose, 1984). Likewise, 'neo-rural' forms of counterurbanisation may be based on countercultural values (Halfacree, 2009) whereby movers seek alternative lifestyles or communities based on, for example, green lifestyles (Halfacree, 2001), gender (Browne, 2009), sexuality (Smith and Holt, 2005) or religious beliefs (Tonts, 2001). Cultural capital, or an ability to engage in the cultural practices associated with these lifestyles, is a strong enabler of both marginal and neo-rural gentrification.

Gentrification is, therefore, multifaceted and diverse, reflecting different social conditions, economic opportunities and the role of various individuals and agencies.

The growth of the service classes and their interests has led to a new 'politics of the rural' (Woods, 2007) that questions what is meant by rurality and how it should be regulated. This has led to conflicts that range:

> *from small-scale disputes over barn conversions, blocked footpaths, street lighting or tree-felling, to large-scale conflicts over new roads, windfarms, waste dumps and major housing developments, the appropriateness of the development to the rural setting and the impact on the rural character of the locality are commonly evoked alongside issues of environmental impact, pollution, noise disturbance, traffic, property devaluation.*
>
> *(Woods, 2007: 884)*

Ideas of rural community and continuity have often been deployed in these disputes. An embattled local farming community might portray itself as defending a disappearing, traditional way of life from urban newcomers (Newby, 1978: 1266). In this context, 'local' people might include a poorly paid farm labourer and a wealthy landlord. Clearly their experience and needs are very different, but the rhetoric of 'locals' versus 'newcomers' glosses over these inequalities. These kinds of arguments were deployed very clearly by the UK's Countryside Alliance in its opposition to a ban on hunting with hounds (Woods, 2003). Using the slogan 'Fight Prejudice, Fight the Ban', the Countryside Alliance portrayed people who lived in the countryside as a social minority who were being unfairly treated by an urban majority who were preventing them from following their cultural traditions. The 'Liberty and Livelihood' protest in London was supported by over 40,000 people and presented 'the view of a somewhat timeless, highly valued and all-embracing country life that needs to be preserved at all costs from the ravages of urbanism' (Cloke 2013: 721). Hunting was seen as central to rurality and being a rural person; yet one protestor noted this hid more divergent views: 'Some of the rural issues we agree with but we are against hunting, this is a political movement, these people are not poor farm labourers they are wealthy people who want to protect a cruel sport' (quoted in BBC 2002).

There is, of course, a geography to these changes and conflicts (Murdoch, 2003). Accessible rural areas are more likely to be colonised by new middle classes, while other, remoter places are still dominated by landowning elites. Seen in this way, social differences and conflicts are products of social wider social and economic change rather than status as local or newcomer *per se*.

8.4 Social Capital and Social Change

Social, economic and cultural changes are closely bound up with each other. When considering these relationships, it is helpful to draw on Pierre Bourdieu's (1984) work on capital. Although capital is often thought of in economic terms, Bourdieu identifies four inter-related forms of capital:[1]

- economic capital – material wealth;
- social capital – social networks and contacts;
- cultural capital – knowledge and skill; and
- symbolic capital – how other forms of capital are represented symbolically (and legitimately).

Particular importance has been placed on the relationship between economic and social capital. If economic capital relates to the wealth of an individual or locality, then social capital refers to the richness of the social relationships between people (Falk and Kilpatrick, 2000). It has been defined in various ways but, broadly, refers to social networks and the ways that they enable beneficial, collective action (Woolcock, 1998; Falk and Kilpatrick, 2000). Robert Putnam (2000: 664–665) describes social capital as the 'features of social life – networks, norms and trust – that enable participants to act together to pursue shared objectives'.

Crucially, different forms of capital can be transferred between each other. Thus, by developing strong social capital, it is also possible to generate economic wealth. For example, developing closer community connections might, in turn, lead to the development of a community-run shop that provides an outlet for local businesses to sell their produce. Running the shop relies on strong cultural capital (education and training), and the shop itself might become an important piece of pride or symbolic capital in the community. Consequently, many rural development schemes have focused on developing social as well as economic capital (Chapter 10).

Box 8.2 Sport and Social Capital in Rural Australia

In rural Australia competitive sport plays an important role in shaping social relationships in rural localities (Figure 8.5). In a study of the Northern Wheatbelt in Western Australia, Matthew Tonts (2005) estimated that 63% of adults played sport (compared to 30% nationally), with others involved as supporters or administrators. Sport was valued for its social interaction:

> *In a lot of ways sport is a sort of glue that keeps us all together. Most people follow it or are involved in one way or another, and I think this means that we get to know each other better. Lots of benefits flow on from that, the main one being that I think we are generally a pretty tight knit and proud community.*
>
> *(male interviewee in Tonts, 2005: 143)*

It was felt that sport drew farmers into towns from remote farms, imbued civic pride and reflected the values of a town. Some interviews reported what Putnam (2000) calls 'bridging capital' that is outward looking and seeks to draw people into social networks. One interviewee said,

> *It really doesn't matter too much about your background. Sport really is a great leveller, so there are people from all walks of life here: farmers, council workers, teachers, the whole lot really.*
>
> *(Female interviewee, aged 40–50, Mullewa)*

Some felt this was particularly pertinent in overcoming racial divisions:

FIGURE 8.5 A cricket match at Dumbleyung in the Western Australian Wheatbelt. Post for Australian rules football can be seen in the background.
Source: Les Everett.

I never really felt like an outsider. I was always part of the team and I reckon it made me a lot of white mates. Things were always easier for me than my relatives, because I played footy. If you didn't, you're just another Blackfella.

(Male interviewee, aged 30–40, Three Springs; Tonts, 2005: 144)

The strength of this social capital translated into economic capital. Some interviewees suggested that being a good player or volunteer meant that people were more likely to use their business or employ them (see also Liepins, 2000a). There was a tendency to 'buy local' or use sports clubs to draw upon voluntary labour.

These forms of social capital have been threatened by out-migration caused by declining farm employment. Willis (2004) notes how the recession in New Zealand's dairy industry led to a loss of rugby players, the closure of rugby clubs and a decline in local identity. Similarly, Tonts and Atherley (2005) show that participation in sport has declined in rural Australian towns that have suffered economic decline. This, in turn, led to further economic decline, with shop owners and bar managers reporting reduced takings on (former) match days, and a waning sense of community.

Box 8.3 Service Provision: The Case of Rural Policing

Rural services are provided in four ways.

State or Public Services: These are services provided by governments – either local or national – and paid for through taxes. Examples include schools, emergency services and, in some countries, health and transport services. State provision aims to level out inequalities by, for example, maintaining services in areas that would not be profitable. The extent to which the state provides services reflects political ideology – in Soviet countries, for example, the state provided community services, such as village halls, as well as health and welfare (Wilson and Klages, 2001).

In many capitalist countries, state services have declined for two reasons. First, austerity and drives towards economic efficiency have meant that governments have rationalised, reduced or privatised services. As the state has withdrawn or 'rolled' back from providing services, governments have become enablers rather than providers of services. In a rural context, village schools have been closed or merged (Ribchester and Edwards, 1999), housing has been built by the third sector (Yarwood, 2002b), public utilities have been privatised (Bayliss, 2003) or services, such as the police, have been centralised in urban locations to improve efficiency (Yarwood, 2007a).

Second, state services are provided through neoliberal free-market principles that emphasis consumer choice. In the case of education, parents can use league tables and quality reports to choose schools for their children, with consequences for the sustainability of small village schools (Walker and Clark, 2010).

The Private Sector: Services are provided by the private sector to make profit and include shops, pubs or banks. The private sector has become increasingly important in neoliberal economies, but, given sparse rural populations and travel costs, services are likely to be more expensive to operate and offer customers less choice than larger, urban branches. Competition from urban stores, especially supermarkets, have made many private services unviable in rural places, leading to their closure. Where services remain, they have often been reinvented for elite customers or tourists. By way of example, 13 rural pubs were shutting every week in the UK in 2012 (Muir, 2012), although there has been a reported rise in pubs in tourist areas such as the Highlands of Scotland, the Lake District or Ceredigion in Wales (Davies and Partington, 2018). These are less likely to cater for local clients and, instead, have focused on catering and accommodation for tourists.

Voluntary or Third Sector: Services are provided by volunteers or charities. It is a remarkably complex sector that ranges from grassroots organisations, such as village hall committees, to corporate non-profit organisations, such as housing associations (Fyfe and Milligan, 2003). Voluntary provision can be sensitive to community needs and allow local people a strong degree of control, but it is also piecemeal, likely to favour those with the time, confidence and wealth to contribute. Consequently, it is often viewed as a 'sticking plaster' that addresses the symptoms rather than cause of service decline.

Partnerships: Services can be delivered through partnerships that formally combine the public, private and voluntary sectors (Edwards et al., 2001). Partnerships attempt to combine the best features of each sector, but questions have been raised about their accountability and the power relationships within partnerships (Cheshire et al., 2006; Goodwin, 1998). Their growth represents a further shift away from the state provision of services and towards new ways of governing the country.

These different approaches can be seen in the provision of rural policing, which is caught between demands for efficiency on the one hand and a desire for greater community accountability on the other (Mawby and Yarwood, 2010). Rationalisation has driven the location of police resources towards centralised, urban locations (where crime is higher) but, at the same time, has been countered by demands for visibility and accountability (Table 8.1) (Yarwood, 2003, 2021, 2022). Although crime rates are lower in the countryside, it is widely recognised that there is also a need to take rurality into account in operational policing (Yarwood and Gardner, 2000; Wooff, 2015).

TABLE 8.1 Staffing of Police Forces in England and Wales by Rurality

Rurality of Forces[2]	No. of Forces	No. of FTE Constables	FTE Constables/ 100,000 Population, 2014	No. of PCSOs	PCSOs as a % of FTE Constables	No of Specials	Specials as % of FTE Constables	Crimes per 1,000 Population
Most Rural	4	5,086	141	747	15	727	15	45
Less Rural	9	13,826	139	1,673	12	2,348	17	49
Middling	14	30,801	138	3,508	12	4,973	16	57
More Urban	11	28,219	157	3,255	11	3,500	12	58
Most Urban	5	49,917	233	3,091	7	6,211	12	60
England and Wales	43	127,849	175	12,274	10	17,759	14	61

Source: Aust and Simmons (2002); Her Majesty's Inspectorate of Constabularies (2015)

One solution has been to establish voluntary policing schemes, notably in the form of farm watch or neighbourhood watch (Figure 8.6). These are established and supported by the police but rely on neighbours to look

FIGURE 8.6 Voluntary policing in the form of Farm Watch in Devon, UK.

out for each other's property and report suspicious activity to the police. Many schemes operate in areas where crime rates are already low, and so it is hard to gauge their impact on crime (Yarwood, 2012), but schemes seem to increase feelings of security and improve relations between the police and the public (Yarwood and Edwards, 1995).

Neighbourhood watch tends to rely on existing social capital rather than develop it. Further, given the territorial nature of NW and the requirement to report 'suspicious people' or 'activities', NW represents a form of bonding capital (Yarwood, 2000: 50). In one rural scheme, residents worked with the police to identify local problems and implement solutions to them. Young people were seen as the main instigators, and a series of situational crime prevention measures were put in place to resolve them (Yarwood, 2010). While many of these solutions were fairly innocuous, such as competitions to pick up litter or hosing down public areas to make them more attractive, it reflects a desire to maintain a hegemonic image of rurality through the exclusion of certain groups or activities (Yarwood, 2005b). In the most extreme example, albeit in an urban setting, an unarmed, young black man, Trayvon Martin, was killed by a NW member when walking home from a shop (Yarwood, 2020a).

Equally, a decline in economic capital may also lead to job losses, out-migration, the loss of previous social networks and a corresponding decline in social capital (Box 8.2). A decline in rural services is not only of economic concern (Higgs, 2003) but has implications for social interactions and the development of social capital (Alston, 2002). Thus, schools are valued beyond their educational function because they provide buildings for community use (for example, fields for a village fete), a social point where people mix (as parents congregate outside the school gate) and help to retain young families in rural places (Walker and Clark, 2010; Witten et al., 2003; Ribchester and Edwards, 1999; Kovács, 2012). More generally, as services have declined, there has been a growing reliance on the voluntary sector to fill the gaps left by state and private provision (Box 8.3), something that both reflects and develops social capital.

That said, social capital has a dark side. Putnam's ideas have been criticised for portraying a harmonious ideal, perhaps steeped in civic institutions that have lost their relevance, and denying 'civil society as an arena of social contestation' (Amin, 1996: 327). Indeed, some services, such as churches, the post office, local police constables, the local store, the pub or the village doctor, are valued for their association with a traditional image of village life as much as their economic value. People also make valuable contributions to society without feeling a need to participate in formal social networks (MacKian, 1995).

Putnam himself recognised that 'bonding capital' is problematic. This refers to forms of social capital that can favour certain groups and exclude others. Returning to Tonts' (2005) example of rural sport, some indigenous players noted that they felt included when they played sport, but race was a barrier to participating in other social networks. Thus, indigenous Australians are under-represented in shire councils, local partnerships or community groups. Bonding capital shows that social capital can be fractured along the lines of class, race, gender and perceived status.

8.5 New Communities

As these previous sections suggest, social change is multifaceted and uneven. It is important to consider how economic change, cultural differences and new social networks alter settlements and the lives of people living in them. Ruth Liepins (2000a, 2000b) returned to the idea of community (Chapter 1) to understand this complexity. She reconceptualised community as a social phenomenon that holds people together through practices, space and meanings (Figure 8.7). Drawing on three case studies (two in Australia and one in New Zealand), she traced how community still has importance in the rhetoric and organisation of rural society, specifically in the following ways:

- meanings of community were expressed through rural landscapes and agriculture, as well as a recognition of different groups of people;
- practices of community were undertaken in both regular (such as playing sport) and infrequent activities (e.g. annual community events). These provided ways of maintaining social connections as well as enforcing or challenging norms of gender, class and sexuality;
- space such as shops, halls and parks, which reflected and shaped community identity. (Table 8.2).

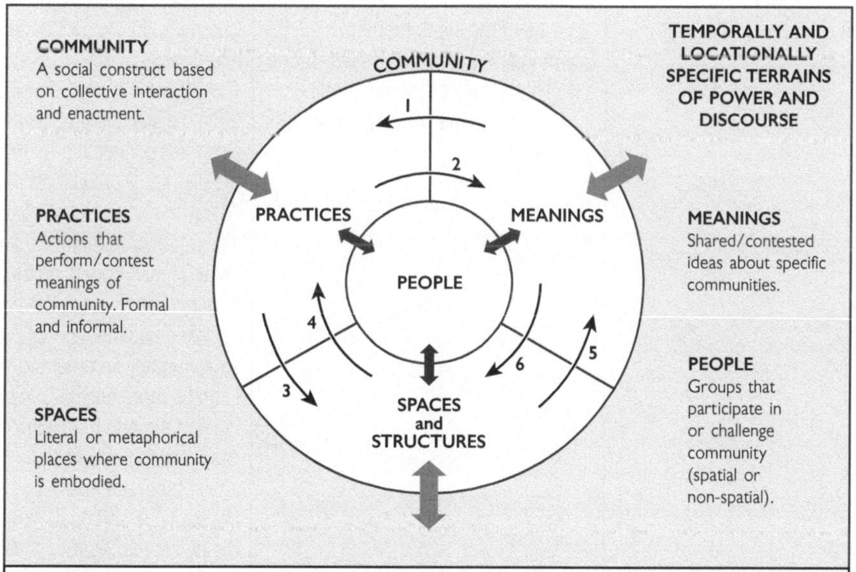

1	MEANINGS	legitimate practices
2	PRACTICES	enable the circulation and challenging of meanings
3	PRACTICES	occur in spaces and through structures, and shape those spaces and structures
4	SPACES and STRUCTURES	affect how practices can occur
5	SPACES and STRUCTURES	enable the materialization of meanings
6	MEANINGS	are embodied in spaces and structures

FIGURE 8.7 Liepins' (2000a) conceptualisation of community.

TABLE 8.2 Residents' Understandings of Community

Meanings	**Practices**	**Spaces**
Topography, history, functions, fragmented	Face-to-face meetings, clubs and societies, annual events	Natural features, services, meeting places
[The community] is their local centre. [Going] to butcher, the baker, and the milk bar and that sort of thing, that nucleus is there for people to go down into the town and get what they want. (Newstead: Interview) Specific places for Duaringa are where people can catch up with each other. I suppose you could say [the] post office is like that. I suppose the hotel is like that. They've been long standing buildings in the community where we can meet each other. (Duaringa: Interview 4) You find out about things going on in Duaringa through the Post Office and the school. Going to the Post Office and, there's the school newsletter I would have to say. And word of mouth. I see a fair few people at the school. (Duaringa: Interview 5)	Often we're involved with the organizing [of events] and the food. Oh and the men . . . [like] 'Bruce', he's good on the BBQ. But really the men are mainly into the action . . . [laughter] . . . You know, mud! Your best chance for fundraising with the men comes down to something that involves dirt, oh and beer. (Duaringa, Interview 11) The rugby club's got quite a bit of influence here. It's a major influence in this district actually. We've had cases where umm say like if a farmer wants a single shepherd. One of the first things they'll check out is if he can play rugby. So that's – I suppose that's a bit of clout. Like they wouldn't look at someone who's forty-five years old for a casual hand, you know, if he didn't play rugby. They'd take someone who could play rugby. And one of the stock feeders downtown, I won't mention which one it is. That's one of their credentials – to be able to play football. (Kurow: Interview 11)	For me, I think of the Dawson [River] when I think of Duaringa. Where you just cross the bridge coming out [of the township]. There are certain sections of that where everything's so tranquil . . . The water's there so that you can forget about the drought or whatever. And everyone has to cross the bridge. (Duaringa: Interview 1) I think the community is epitomized by the sports sites . . . the race track, and there's the sports grounds which would be one of the main things for Duaringa. Up there we have not only the racing facility, but there's also the golf club, the rodeo club, the pony club. There's a lot of activity centred up there. It's central for the Bullarama and the Charity Golf Day. (Duaringa: Interview 2) The school is a huge connecting force. I'd say it is probably the strongest connecting feature in the district. (Kurow: Interview 1) I'd say the hotels must do pretty well . . . they're really the focal point. I mean people go there not to get drunk, they go there to socialise and meet people. And as somebody commented to me, I can do more business in half a night in the pub than I can do in a week staying at home. So, you know, there are business transactions and deals done in the hotels too. (Kurow: Interview 3)

Source: Liepins (2000a)

The communities studied by Liepins were based on agricultural economies and practices. Community events, for example, helped to keep farmers 'going' and reinforce town identities as 'farming communities'; planning decisions and local

FIGURE 8.8 Community space in Western Australia.

governance were geared towards supporting agriculture. Yet as farming practices changed and fewer people worked on the land, some residents noticed changes in some community practices:

> *There are less of us [farming households] in the district now and it seems you work harder or get older. We're not involved in the tennis club any more. And younger ones are often going out of Duaringa for their weekends now. That changes the place.*
>
> (*Duaringa: Interview 8*)

Building on the tradition of community study in rural geography, Liepins' work provides a nuanced way of considering the relationship between economic changes, the ways that people live their lives and how these are imagined and reproduced in everyday life.

8.6 Conclusions

The country has undergone profound social transformations, yet it is often portrayed as unchanging and unchangeable. The idea of community is resolutely upheld by many who live in rural places, although the term can often hide as much as it can reveal. It is tempting to see rural society as an outcome of wider processes that work through and across different rural spaces. What is clear, though, is that local places and cultures also affect the ways that change occurs. Rural geography has a complex task in unpacking not only the mechanics of change but also the rhetoric of change and the power relations that often shape or mask how change is contested. Although the countryside continues to be portrayed as harmonious and problem-free, change exposes fault lines and conflicts in rural society, as the following section examines.

Notes

1 It might also be possible to add forms of capital, including environmental capital (Wackernagel and Rees, 1997) or enthusiastic capital (Yarwood and Charlton, 2009) to name but two.

2 Based on Aust and Simmons' (2002) classification of rural police forces.

Part III Contests

The previous section used Halfacree's (2007) threefold model of rurality to examine some of the ways that rural areas have changed. It traced how agricultural and economic restructuring has transformed the countryside; the ways that the countryside has been imagined through various media and how the social structures of the country have undergone profound change. These transformations have not always been harmonious. Rural change has been characterised by various conflicts, as the titles of many rural textbooks imply (Cloke and Little, 1997; Yarwood, 2002a; Robinson, 1990; Macnaghten and Urry, 1998). The following section examines how rural localities are contested in different ways and what these tensions reveal about rural society.

DOI: 10.4324/9780429448966-11

CHAPTER

9

Poverty and Social Exclusion

9.1 Introduction

The country has undergone wide-ranging economic, social and cultural change. These transformations have benefited some people, but others remain economically and politically marginalised. While poverty is often associated with inner cities, 80% (c. 610 million) of the world's poorest people – defined as having less than $1.25 a day – live in rural areas (de la O Campos et al., 2018). Globally, the rural poverty rate is 19.2% in rural areas, which is three times more than in cities (United Nations, 2020b). In the European Union, the highest risk of poverty and social exclusion is found in rural areas, especially in accession countries (Figure 9.1). In the USA, the poverty rate was 16.4% in non-metropolitan areas and 12.9% in metropolitan ones (United States Department of Agriculture, 2019; especially amongst women, single parents and ethnic minorities (Tickamyer, 2006; Tickamyer et al., 2017). Rates of poverty are also high in rural Australia (Table 9.1), especially in more remote regions (Alston, 2000): in 2015–2016, the net household worth was, on average, 29% lower for people living outside major cities (National Rural Health Alliance, 2017). In a rural context, poor services and a lack of transport also contribute to a wider 'jigsaw' of deprivation (Shaw, 1979).

Yet despite this evidence, poverty often remains hidden in many rural localities (Milbourne, 2016). The rich and poor can live side-by-side, making it difficult to identify where poverty occurs and, consequently, how to deal with it using spatial policies (Liu et al., 2017; Pacione, 1995). Idyllic notions of rurality continue to deny the existence of poverty in the country, meaning that it is rarely prioritised by policy-makers (Cloke et al., 1995a; Woodward, 1996). As this chapter shows, it is important to understand the spatial distribution of poverty, how it is perceived by policy-makers and, perhaps most significantly, how it is experienced by those living with it.

DOI: 10.4324/9780429448966-12

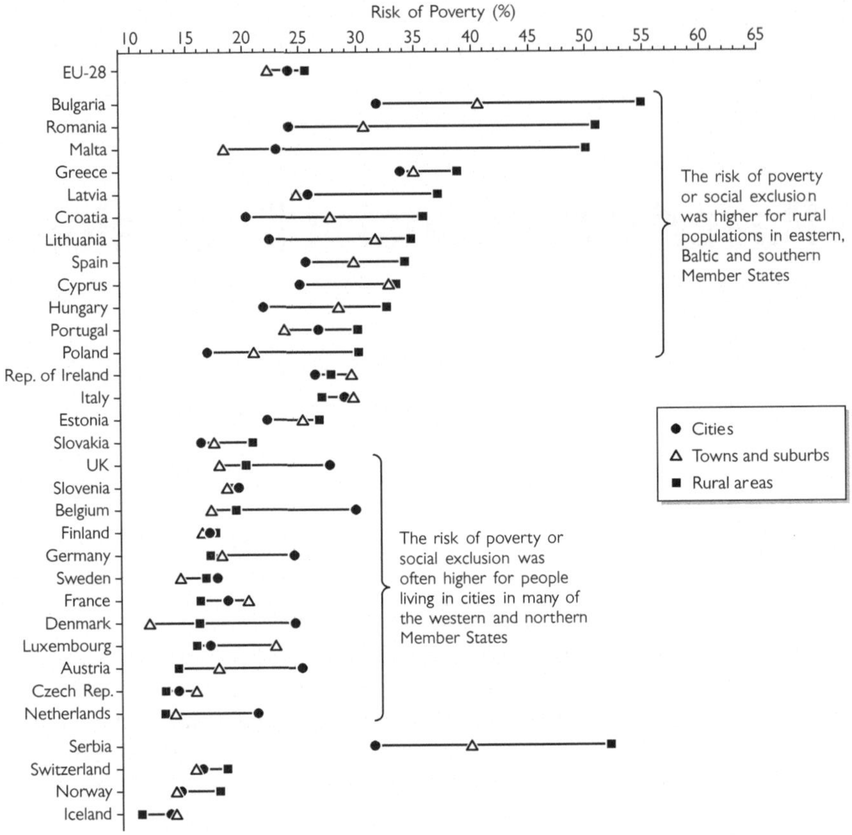

FIGURE 9.1 Rural and urban poverty in the European Union.
Source: Bertolini (2019)

TABLE 9.1 Mean Income and Wealth in Australia (2015–2016)

	In Capital Cities	Outside Capital Cities	Percentage Difference
Disposable household income	$1,072	$880	18%
Net household worth	$1,033,000	$737,000	29%

Source: National Rural Health Alliance (2017)

9.2 Defining Poverty and Exclusion

The World Bank's (2018) measure of $1.90 a day is widely used to chart global poverty (United Nations, 2020b). This kind of *absolute* measure provides a helpful objective benchmark with which to quantify, map and track poverty (Smith et al., 2010). According to this measure, the percentage of the world's population living in poverty fell from 36% to 10% between 1990 and 2015 (United Nations, 2020b). But, during the same period, financial inequalities increased in many countries.

This indicates a need to consider poverty *relative* 'to the local community or the wider society or nation to which an individual, family or group belongs' (Townsend, 1987: 125). As income data can be skewed by extremely high or low figures, median values are a helpful way of examining relative wealth in relation to a 'typical' income. Hence the EU defines household poverty as income below 60% of a median income (Milbourne, 2016) (Table 9.1). This said, deprivation is more than a lack of income (Cloke et al., 1995a) and refers to the inability to meet basic human needs (Shucksmith, 2016). The United Nations (1995) therefore refers to absolute poverty as:

> *a condition characterised by severe deprivation of basic human needs, including food, safe drinking water, sanitation facilities, health, shelter, education and information. It depends not only on income but also on access to services.*

And more widely as:

> *lack of income and productive resources to ensure sustainable livelihoods; hunger and malnutrition; ill health; limited or lack of access to education and other basic services; increased morbidity and mortality from illness; homelessness and inadequate housing; unsafe environments and social discrimination and exclusion. It is also characterised by lack of participation in decision making and in civil, social and cultural life. It occurs in all countries: as mass poverty in many developing countries, pockets of poverty amid wealth in developed countries, loss of livelihoods as a result of economic recession, sudden poverty as a result of disaster or conflict, the poverty of low-wage workers, and the utter destitution of people who fall outside family support systems, social institutions and safety nets.*

Poverty is therefore viewed as an outcome and cause of societal disadvantage that conspires to reduce an individual's life chances. This is reflected in the term social exclusion, which refers to:

> *dynamic, multi-dimensional processes driven by unequal power relationships interacting across four main dimensions – economic, political, social and cultural – and at different levels including individual, household, group, community, country and global levels. It results in a continuum of inclusion/exclusion characterised by unequal access to resources, capabilities and rights.*
>
> *(World Health Organisation, 2020)*

The idea of social exclusion stresses that poverty is not the fault of an individual but, rather, the circumstances they find themselves in. However, as the following section shows, there are also strong cultural dimensions to the way that poverty is (or is not) seen and how it is regarded.

9.3 From Structural to Cultural Exclusion

There is a long-standing belief, particularly in conservative rural heartlands of the USA, that it is possible for people to work themselves out of poverty and, by extension, that they remain in poverty because they are more likely, or more willing, to depend on welfare. As these views are often associated with social or racial stereotyping, they have little traction in academia (Shucksmith, 2016). Nevertheless, they are used to support policies, such as workfare, that are aimed at reducing the state's responsibility to support those in poverty. Tickamyer (2006: 415) points out that conservative views on welfare, combined with a cultural reluctance to use it, mean that 'largest concentrations of poverty coincided with states with lowest benefit levels, the most punitive policies and most corrupt and discriminatory local administrations'. As Box 9.1 demonstrates, these represent a form of 'moral capital' that prioritises self-help over welfare support (Sherman, 2006).

Box 9.1 Poverty in Rural USA

Jennifer Sherman's (2006) study of Golden Valley in California, USA, draws attention to the significance of cultural norms in determining responses to poverty and contributing to the social conditions that maintain it.

Golden Valley had a population of 2,000 people that relied mainly on forestry and logging for employment. In 1991 an injunction was placed on logging to preserve the Northern Spotted Owl, which had been listed as an endangered species in 1990. Sawmills closed and large-scale logging ended, resulting in poor employment prospects for those remaining. In 2000, the poverty rate stood at 24% (with one indicator suggesting 14% of households lacked phones), with many other people moving out of the valley in search of work. Between 1990 and 2000, unemployment rose from 11% to 21% and the number of men in the workforce declined from 56% to 48%. More women entered the workforce but often in part-time, poorly paid jobs. Remarkably, the use of welfare declined from 19% to 9.5%.

Sherman explains that a form of 'moral capital' emerged in Golden Valley that allows 'the poor to create distinctions among themselves' (893) despite a lack of material wealth. Different types of moral capital were supported and reflected in a continuum of activities (Figure 9.2). Forms of subsistence, including hunting, fishing and growing food, were 'almost universally practised' (897) and were considered morally acceptable alternatives or supplements to paid work. At the other end of the scale, living on welfare or illegal activities, especially drug dealing or using, were regarded negatively. Others tapped power cables for illicit electricity or phone lines. Disability, which might be a consequence of an industrial accident, entitled people to Supplemental Security Income, but many chose to work rather than accept this. Others who claimed welfare for

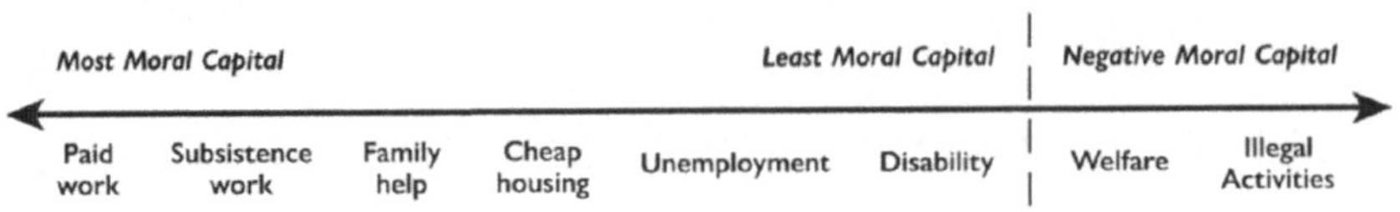

FIGURE 9.2 Moral capital and rural poverty.
Source: Sherman (2006)

disability were viewed with suspicion by those who felt that they could still work.

Drawing on work by Bourdieu (1984), Sherman argues that moral capital is important because it can be converted into social or economic capital – in other words, following moral conventions leads to a higher standing in the community and improved social and economic prospects. Those of high moral standing are more likely to receive informal community help or support in hard times. People with low or negative moral capital were less likely to receive help from the community or offers of employment. Put more simply, moral capital defined and distinguished between the 'deserving' and 'undeserving' poor.

Sherman's study also highlights the significance of understanding local social relations and the importance of cultural norms when studying poverty. It also points to the prevalence of cultural norms that assume poverty can be overcome by individuals and communities. These discourses deflect attention from the wider economic and political factors that create these conditions in the first place. They also promote welfare policies that have been prevalent in the States, and now Europe, that support the restructuring of welfare towards forms of workfare that underline and support ideas of moral capital at national and local levels.

The danger with these viewpoints is that they fail to acknowledge the structural disadvantages that cause and maintain poverty. Core-periphery models argue that rural places are 'left behind' or 'bypassed' by the modern world because they are too isolated or lack the institutions and dynamics to benefit from new rounds of capital (Tickamyer, 2006). This is reflected in a political rhetoric of 'levelling up', where aid and support are given to poorer areas to help them 'catch up' with wealthier ones. The European Union, for example, had policies that targeted 'lagging' rural regions (Terluin, 2003).

The rural poor may also be exploited by capitalists (Tickamyer, 2006). Industrial branch plants are often established in rural areas to make use of a workforce with few opportunities for alternative employment. In these places, wages can be kept low and often go unchallenged due to a lack of trade unions and the footloose nature of industrialisation. Women provide cheap, green labour in rural places, immigrant labour is provided by ethnic and racial minorities and power remains in the hands of elite groups, be they landowners or new middle-class migrants (Woods, 2006). As these powerful people occupy decision-making positions, there is little incentive to make radical changes that may alleviate poverty in the country.

During the 1990s, Paul Cloke and his colleagues (Cloke et al., 1995a, 1997; Cloke and Milbourne, 1992) undertook a major study into rural exclusion in England and Wales. Their work confirmed the range of issues facing people living in the countryside, such as a lack of housing, poor employment prospects, low income and issues of well-being, and recorded similar rates of poverty to those found in earlier studies (McLaughlin, 1986b), albeit with variation between and within different places. Their work also revealed that disadvantage is experienced in different ways. Thus, isolation, which is often used as a measure of deprivation, is a barrier for some people that prevents access to work, services and friends; but for others, it offers a sense of 'splendid isolation' sought in rural life. Their work also revealed that cultural constructions of rurality were interwoven with experiences of poverty (see also Box 9.2). For example a lack of services was often accepted stoically as part and parcel of rural life; and idyllic views of rurality were used to deny the existence of poverty by:

> *those who experience hardship (but will perhaps see this as an acceptable trade-off for the benefit of rural living or will seek for some reason to conceal or underplay the stigmatic acknowledgement of hardship) and for those who do not (and perhaps are anxious to reproduce the culture of an idyll by playing down any hardship that comes to their attention).*
>
> (Cloke and Milbourne, 1992: 359)

Box 9.2 Rural Poverty in the UK

Paul Milbourne's (2004) study of Wiltshire in south-west England reveals how significant levels of poverty occur in the midst of an affluent county. At the time of his study, nearly 12% of the county's population were part of households that received benefits, and, of these people, 37% were over pensionable age, 10% were lone parents and 28% were dependent children. These people were dispersed across the county, with little evidence of concentration. That said, the parishes with the highest levels of poverty were the most isolated and had high levels of material deprivation, indicated by free school meals and lack of car ownership. Poor pensioners were found in every parish. These inequalities were exacerbated by a decline in services and social housing due to low levels of state and private investment. As a result, young people were likely to leave in search of work or better lifestyles. Despite these issues, the planning system favoured landscape preservation rather than social development, helping to maintain social inequalities.

A closer analysis of one village, Patbourne, revealed that 'almost all of the low income benefit households reside in a small estate of social housing situated away from the historical core of the village' (p. 569). This had implications for the imagined and lived social geographies of village life. Poor people lived out of sight of the village centre and were, consequently, not in the consciousness of other, wealthier residents. They were often regarded as peripheral to village life but remained trapped there due to a lack of income, transport and opportunity. Wealthier residents lived their lives across a much greater area, travelling widely to work, shop or socialise outside their home village.

These social divisions caused tensions that were played out in various community groups and organisations. Poorer people felt stigmatised, less able to participate and unwilling to disclose their situation to the wider community. Far from being close-knit, local social structures contributed to the stigmatisation of the poor and led to feelings of exclusion.

Elsewhere, Milbourne (2014) notes that local communities provide an important context for many people in poverty. Despite low incomes, many regard their local places and communities favourably because their location and landscape offered some compensation for material poverty. Sherman (2006) also argues that these forms of social cohesion put pressure on 'the poor to behave in ways that are consistent with local values' (891). Thus, as Milbourne continues, these feelings of community can also hide poverty and contribute to a sense that self-help and self-sufficiency can resolve, or at least mitigate, poverty. These ideas deflect attention away from the need for strong state intervention to resolve poverty.

9.4 Conclusions

Writing nearly 30 years ago, Paul Cloke (1993) argued ominously that rural poverty would remain a prevalent issue in rural places. This was for four reasons. First, solutions to solving poverty have often been partial, relying on local self-help schemes that address symptoms rather than causes of poverty. These are piecemeal and rely on an ideal that caring rural communities can look after themselves through voluntary schemes. These structural disparities require radical solutions rather than sticking-plaster approaches that treat the symptoms rather than the causes of poverty. Second, privatisation has swamped potential solutions. Governments have 'rolled back' the provision of many services, such as housing or transport, and replaced them with private companies that are unlikely to invest in low-profit places. Third, the problems are rooted in the nature of society, meaning some groups are more likely to be disadvantaged due to class difference and a lack of power. Finally, and depressingly, the same problems are reproduced with new cohorts of people. As the next chapter shows, although some changes have been made to the planning and governance of rural areas, social issues take second place in many rural policies.

CHAPTER

10

Policy and Governance

10.1 Introduction

The countryside is complex, diverse and conflicted. There is therefore a need to plan for rural areas and develop policies that address differing concerns (Scott et al., 2019). Rural planning and policy might be viewed as mechanisms for arbitrating between competing stakeholders in a neutral and fair manner. Seen more critically, efforts to plan, govern or regulate the country reveal much about 'the way power has been exercised *within, over, by and about* the rural' (Cheshire, 2016: 494). Rural policies might, for example, promote the country as a place of industrial food production or, instead, a place that is preserved and protected for tourists or wildlife. Or indeed both. These reflect not only wider debates about rurality but the relative power of different stakeholder groups. The farming lobby, for example, has often been very powerful in the formation of rural policies.

A systematic and detailed review of rural policy in specific countries is beyond the scope of this book and would, in any case, become quickly dated. Consequently, this chapter traces broad trends in the governance of rural places and encourages readers to think critically about these issues when considering policies and decision-making in more specific contexts. This approach allows us to not only understand how and why the country has changed but, more significantly, who has the power to effect change. This is pertinent given radical changes in rural policy since the Second World War (Winter, 1996; Curry and Owen, 2009). To that end, this chapter traces shifts in rural policy, changing structures of governance and what this means for rural citizens.

10.2 Changing Rural Policies

Rural politics and planning have traditionally been associated with the production of food and the regulation of land (Winter, 1996). These led to a labyrinthine set of policies that attempted to regulate food production and manage landscapes, whilst paying scant attention to social issues or welfare beyond those associated with farming (Woods, 2006). In response to the economic, social and environmental

DOI: 10.4324/9780429448966-13

impact of intensive farming, policy broadened to encompass a wider range of economic activities and to reflect more holistic ideas of what the countryside was, who lived there and how decisions should be made. Increasing emphasis was placed on forms of endogenous growth (that emerged from within areas) rather than forms of exogenous growth that originated from outside a locality (Table 10.1). These can be summarised as a shift from *agricultural* to *rural* policy (Table 10.2).

TABLE 10.1 Exogenous and Endogenous Development

	Exogenous	**Endogenous**
Key Principle	Economics of scale and concentration	Specific resources of an area hold the key to its development
Dynamic Force	Urban growth poles; main development forces come from outside rural areas	Local initiative and enterprise
Function of Rural Areas	Food and primary production	Diverse service economies
Major Rural Area Problems	Low productivity	Limited capacity of social groups to participate
Focus of Rural Development	Agricultural industrialisation; labour and capital mobility	Capacity building; overcoming social exclusion

Source: Gkartzois and Lowe (2019)

Rural Policy in the European Union

The Common Agricultural Policy (CAP) was introduced in 1962 by the European Economic Community (EEC, later the European Union) as a price support policy for farmers (see Chapter 5). As time went by questions were raised about the amount of support given to farming in relation to other economic sectors (Figure 10.1). In 1984, CAP accounted for 73% of the EEC's budget, although farmers are only 3% of the EU's population and contribute 6% of GDP (An Oifig Buiséid Pharlaiminteach/Parliamentary Budget Office, 2018).

TABLE 10.2 Shifts in Rural Development Policies

Field	**Shift From**	**Towards**
Focus of Rural Development	**Agricultural industrialisation; labour and capital mobility**	**Capacity building; overcoming social exclusion**
General development measures	Exogenous	Endogenous
Agricultural policy	Productivist	Post-productivist
Coverage of policy	Sectoral	Territorial
Governance	Top-down	Bottom-up and partnership

Source: Terluin (2001)

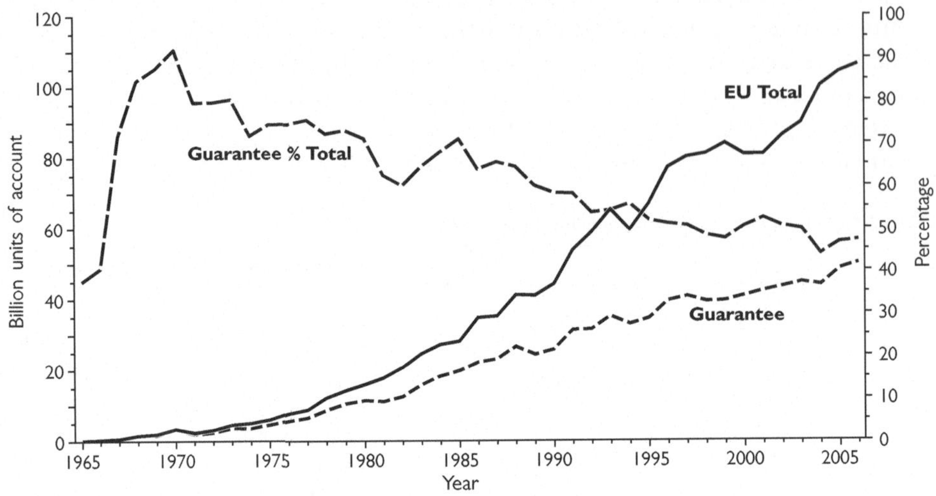

FIGURE 10.1 Total EU spending and CAP guarantee spending, 1965–2000.
Source: Ackrill (2008)

Additionally, some 'lagging' EU regions were the focus of successive funding rounds that aimed to stimulate economic development and create jobs. Between 1989 and 1999, these included 'Objective 5b' funding that targeted rural regions with below average levels of economic development (Ward and Mcnicholas, 1998a). The aim was to stimulate development from within the region (endogenous) by supporting small and medium enterprises that promoted economic diversity. After 1999, rural areas were not specifically the focus of structural funding but were included as part of programmes that offer regional support to less-developed regions, especially in Eastern Europe (Figure 10.2).

Since 1991, a second, smaller strand of funding, known as LEADER, has funded small-scale, endogenous economic and social developments in rural places (see Box 10.1).

In 1999, rural development was established as a 'second pillar' of CAP (Ackrill, 2008) with the aim of 'achieving balanced territorial development of rural economies and communities, including the creation and maintenance of employment' (European Commission 2022). This is delivered through national Rural Development Programmes (RDPs) written by member states to meet their specific national needs. RDPs are in place for seven-year periods and allow countries to draw upon a 'menu' of support packages tailored towards specific types of social, economic, agricultural and environmental development. Some RDPs cover a whole country, other states have chosen to implement several regional RDPs and others have adopted a hybrid model with regional RDPs supporting a national plan (European Commission 2022).

Despite these changes, there is evidence that CAP continues to favour large farms (Kiryluk-Dryjska et al., 2020), and there has been limited impact on rural development or rural populations (Lillemets et al., 2022). Although there were positive effects on employment, there is less evidence for diversification, regional cohesion and civil participation. Questions also remain about the impact of CAP on

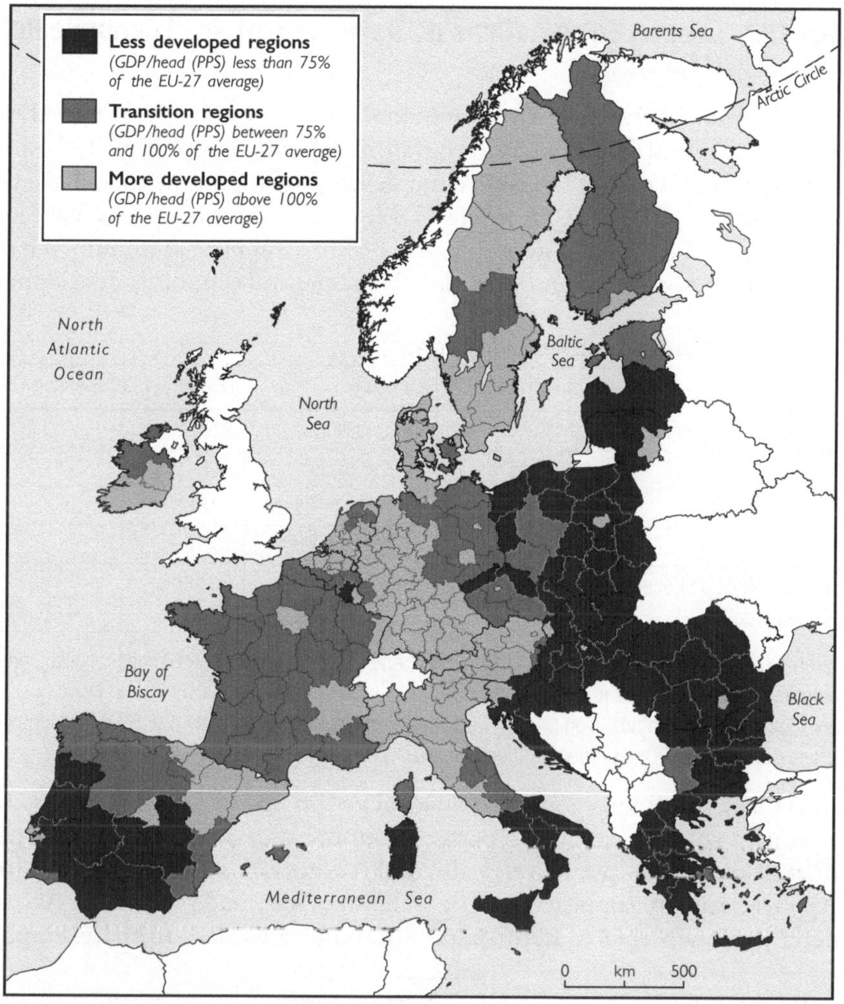

FIGURE 10.2 EU development regions.
Source: European Commission, 2023

generational change and gender equality. Despite this, CAP remains 'the most expensive and extensive of the EU common policies' (p. 298), with a budget of €99.587 billion between 2014 and 2020 (Andersson et al., 2017).

Lapping and Scott (2019) also record that development programmes in the USA, Canada and Japan had been dominated by agricultural rather than rural issues. In the USA, the federal structure has contributed to a rather piecemeal approach to rural policy that relies on local governments and other actors to plan and implement how rural areas are managed. There is, for example, no national land use planning policy in the USA, with decisions and policies negotiated at the local level. Although the US Department of Agriculture provides national support for agriculture and, to a lesser extent, rural development, rural policies have largely been developed at a local level. These have been endogenous in their approach, with varying degrees of success.

Box 10.1 Liaison Entre Actions de Development de l'Economie Rural (LEADER)

Liaison Entre Actions de Development de l'Economie Rural (LEADER) was an EU rural development programme that was piloted in 1991 and is now an integral part of the EU's rural development policy. There have been five rounds of LEADER funding (Table 10.3), and while there have been some changes, the consistent aim has been to support community-based, innovative actions to promote sustainable social and economic development.

TABLE 10.3 Rounds of the LEADER Programme

	Years Active	Funding	Number of Local Action Groups
LEADER	1992–1994	€442 million	217
LEADER 2	1994–1999	€1,755 million	906
LEADER Plus	2000–2006	€2,105.1 million	1,153
LEADER*	2007–2013	€5,903.4 million	2,416
Community-Led Local Development (CLLD)*	2014–2020	c. €5 billion	c. 2,800

*From 2020, the programme was 'mainstreamed' and administered as part of national development programmes both urban and rural places.

Sources: European Commission (2006); European Rural Network Assembly (2015); Publications Office of the EU (2020); European Parliament (2020)

LEADER funding is governed by local action groups (LAGs). LAGs cover a defined rural territory in a partner country and are comprised of a panel drawn from the public, private and voluntary sectors (Figure 10.3). Applications to support specific schemes are made to the LAG that determines whether or not funding should be awarded. Initially, emphasis

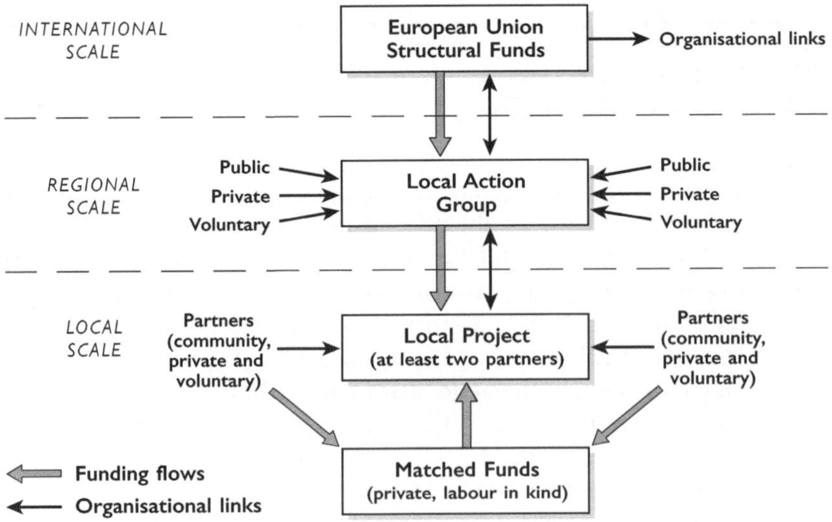

FIGURE 10.3 Partnership working and LEADER.

was placed on providing match-funding for schemes and supporting collaboration between different local actors, but although these are still valued and encouraged, these are no longer requirements. The first three rounds of LEADER had their own EU funding stream, but since 2007, LEADER has been 'mainstreamed' by its incorporation into wider EU structural funds (Dax et al., 2016). In addition, LEADER has expanded to cover coastal and urban areas, thereby tapping into other funding streams, in a programme now referred to as Community-Led Local Development (CLLD). LAGs continue to make decisions about how this funding is spent, but in rural places, this is more closely aligned to national rural development plans drawn up as part of the second pillar of CAP.

Examples of LEADER-funded projects include training for hedge laying in Belgium; dry-stone wall building in Scotland; energy production from olive stones in Greece; numerous farm diversification schemes; manufacturing shinty sticks in Scotland; a village celebration in Portugal; the marketing of local produce in the Montiferru region of Sardinia; and a school of Byzantine art in Greece. These schemes are small-scale and draw upon local resources, skills and traditions to generate income and opportunity.

The programme's longevity is indicative of its popularity and ability to galvanise local support and entrepreneurship. The establishment of LAGs reflects a partnership approach to development that has been praised for improving the capacity of local communities to address local issues and fostering local social and economic connections to enable local development (Scott, 2004).

Despite these successes, LEADER has been criticised for a lack of diversity on LAGs, limited innovation and a lack of transnational cooperation (Storey, 1999, García and Valverde, 2009; Shucksmith, 2000; Shortall, 2008). Questions also remain about the significance of LEADER, when compared to the amount of money spent on agricultural support (Ray, 2000). Others have also questioned whether 'mainstreaming' LEADER has reduced its capacity for bottom-up action as, increasingly, it is required to fit into national development strategies (Dax et al., 2016).

LEADER highlights many of the shifts that have occurred in rural policy, as well as some of the limitations of community-based approaches to tackling social and economic inequalities.

10.3 Government, Governance and Localism

As the example of LEADER implies, there has been a shift in the way that decisions are made in and about rural places. Broadly, this has seen a shift from top-down decision-making by a political authority toward more bottom-up or partnership-based approaches that attempt to enrol a wide range of people and organisations in decision-making (Figure 10.4). Some commentators have suggested that this reflects a shift from govern*ment* to gover*nance* (Goodwin, 1998; Stoker, 1998; Cheshire et al., 2006; Edwards et al., 2001) and a greater emphasis on active citizenship solve rural problems (Box 10.2). Government refers to decision-making through formal institutional structures, such as county or shire councils. By contrast, governance is a much broader concept that refers to the way that governmental and non-governmental organisations work together to make decisions and manage the country.

In the top-down approach, decisions are driven by an authority such as a national or local government, which can often appear centralised and remote from rural places (Figure 10.4). The aim is to take a strategic approach to managing the countryside, perhaps treating it as a national asset to produce food, and to decide how particular areas might be managed in order to meet these objectives. The advantage of this top-down, government-led approach is that it allows planners to think strategically. Villages, for example, might be classed as part of a settlement hierarchy that determines where housing or services are allocated.

This is not to say that top-down approaches are dictatorial or undemocratic, as governments are made accountable through elections, formal channels of communication, consultations and planning enquiries that offer citizens opportunities to have their voices heard about decisions may affect them. Yet there are concerns that citizens ought to be more closely engaged in decision-making rather than simply voting for representatives to make decisions on their behalf. In 1969 Sherry Arnstein (1969: 216) argued that 'the idea of citizen participation is a little like eating spinach: no one is against it in principle because it is good for you' but, in practice, there had been little appetite to involve citizens in planning. She devised a now-famous 'ladder of participation' (Figure 10.5) that identifies different levels of participation. Consultation and appeasement are tokenistic, doing little to promote or

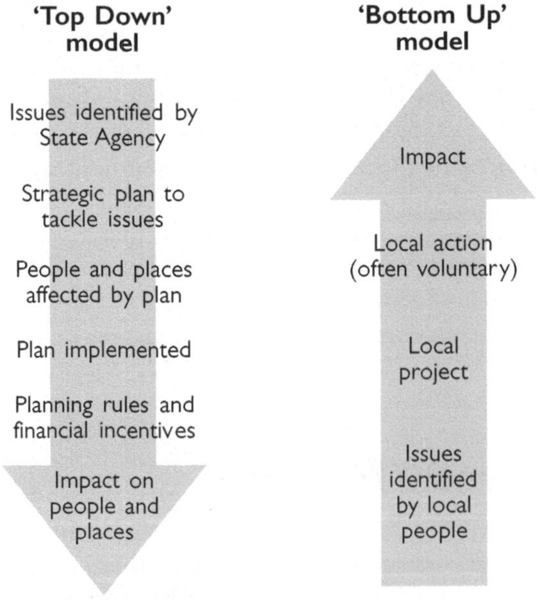

FIGURE 10.4 Models of decision-making.

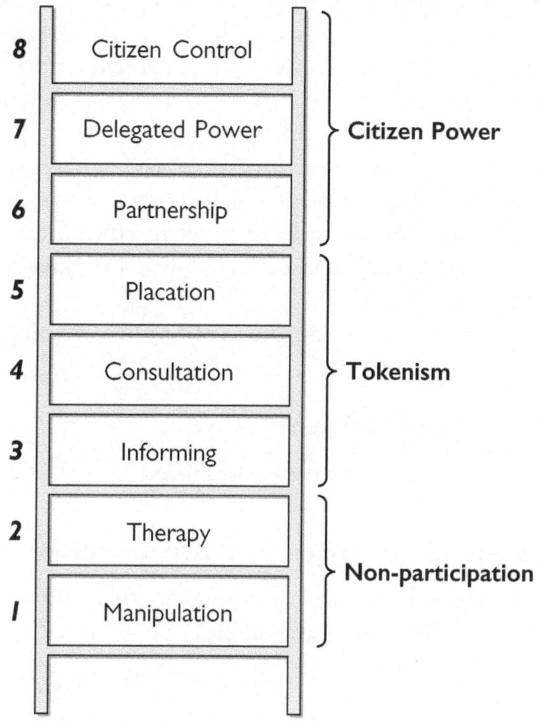

FIGURE 10.5 Ladder of participation.
Source: Arnstein (1969)

respond to citizen power. Bottom-up approaches, by contrast, aim to empower local people to identify and resolve local issues. Around the same time, the Skeffington Committee (1969 [2013]) in the UK argued for stronger community participation in planning, reasoning that planners should work *with* rather than *in spite of* communities.

From the 1980s onwards, there have been stronger efforts to involve local people in the governing of their areas. Rural communities are better placed than urban ones to undertake bottom-up planning because they can draw upon existing local networks and governance structures. For example, in the United Kingdom, parish or community councils are locally elected bodies that have the potential to co-ordinate local action. One example of this was seen in Village Appraisal schemes in the 1980s that encouraged local communities to identify what was important in their village and develop plans for retaining or developing these features (Moseley, 1997). Drawing on support from parish councils, Village Appraisals used questionnaires, parish maps and community projects to encourage local residents to identify what was important to them in their village (Hughes, 1993). The issues identified by appraisals were often small-scale – such as improving a local sports field, clearing waste ground, improving the environment, utilising community buildings and so on – but were important to local people and could be acted on by volunteers.

Village Appraisals helped some communities identify local issues and implement approaches to solve them and their success led to many planning authorities formally adopting Parish Plans in the early 2000s as a way of consulting with local communities and identifying how to meet their needs (Owen and Moseley, 2003; Gallent et al., 2008b). More recently, Neighbourhood Plans have been adopted in order to give local residents a formal role in determining how and where growth occurs in their communities (Box 10.2). Bottom-up approaches have also been used by a range of different agencies to provide housing and services in rural areas (Yarwood, 2005b; Salvia and Quaranta, 2017; Gallent et al., 2008b; Devine-Wright et al., 2019; Edwards et al., 2001; Pemberton (2019).

Malcom Moseley (2003) suggested that community-based development encourages:

1. **local diversity** – different places require different solutions;
2. **solutions to interlocking problems** – through integrated and partnership approaches at local levels;
3. **local knowledge and action** – to identify and act on particular problems;
4. **adding value to local resources** – local trade and action to multiply economic development, which may also be more sustainable;
5. **anti-globalisation** – to create local niche markets for products and people.

Despite these advantages, bottom-up development remains limited in its scope. While it can address small-scale issues that are important to local people, such as how a playing field is managed, these achievements do little to address deep-seated issues that continue to cause inequality in rural places (Vera-Toscano et al., 2020). It has long been described as a 'sticking plaster' (McLaughlin, 1987) approach to rural development, addressing symptoms rather than causes of deprivation. It remains in favour because it is cheap, especially when relying on voluntary effort, and allows governments to shift blame onto rural communities – if they are unwilling to take action, then continuing problems can be viewed as their own making (Cheshire, 2016; Desforges et al., 2005). Communities are therefore seen as both the cause and solution of rural problems (Lockie et al., 2006): the result is a patchwork of services that reflects willingness or capacity to engage in local politics rather than meeting particular needs (Box 8.5) (Desforges et al., 2005).

Box 10.2 From Town and Country Planning to Spatial Planning

In response to food shortages during the Second World War, the 1947 Agriculture Act and the 1947 Town and Country Planning Act in the UK were the twin drivers that defined and designated the countryside as primarily a place to grow food. While the former propelled farming towards intensive production, the latter placed strict controls on built development to ensure that land was reserved mainly for agriculture. While farming flourished, a 'no development' ethic (Curry and Owen,

2009) stifled social and economic development in rural places and, at best, confined it to select 'key settlements' marked for growth (Cloke, 1983).

The 1947 Town and Country Planning Act placed the responsibility for planning in the hands of county councils that were required to produce Development Plans that laid out how land would be used over a 20-year period. These were enforced through the concept of Planning Permission that required landowners to seek consent from local authorities before making any material changes their property. This gave authorities some control as to where houses were built, what land was reserved for farming, where roads could be laid and which land was reserved for conservation or recreation. Significantly, planning permission did not apply to many agricultural activities.

In 1995, the Conservative government published 'Rural England: a nation committed to a living countryside', the first rural white paper for over 30 years, which was followed, rather like two delayed buses arriving at a rural bus stop at the same time, by Labour's rural white paper 'Our Countryside: the Future. A Fair Deal for Rural England'. While there were differences in political emphasis between the two, both papers recognised a need for rural, rather than simply agricultural development, and a desire for rural communities to work in partnership with various agencies to secure a living/fair countryside (Murdoch, 1997; Lowe and Ward, 2001; Hodge, 1996). With the desire to empower rural communities (however defined), there was also an expectation that they could and should solve difficulties in their areas.

Some of these ideas were incorporated into the planning system. The 2004 Planning and Compulsory Purchase Act introduced the concept of spatial planning which attempted to broaden planning beyond development control and land use to embrace a more holistic view of countryside (Gallent et al., 2008a) (Table 10.4).

TABLE 10.4 Key Shifts in the UK Planning Policy

Land-Use Planning	**Spatial Planning**
Narrow scope (development control)	Broad scope (spatial governance)
Tight boundaries (simple)	Loose boundaries (complex: who is in charge?)
Local-authority-driven	Collaborative
Topic-based	Integrated/spatial
Local-authority-owned	Owned through partnership
Planning peripheral Planners not always viewed as central to local authority activity	Planning central Planners more closely engaged
Lacks vision/outcome-led	Shared vision
Land-use objectives	Diverse objectives

Land-use policy	Multiple policy tools Non-site-based policies
Hierarchical	Integrated
Legal	Project-managed (flexible)
Consultation	Engagement
Limited inputs	Building consensus
Limited monitoring	Continual review
Narrow mechanisms (development control)	Broad mechanisms (range of agencies)
Planning outputs (what gets built)	Broad outcomes (e.g., sustainable development)

Source: Gallent et al. (2008a)

These reforms were supported by the 2011 Localism Act, which, amongst other things, introduced the concept of neighbourhood planning and offered opportunities for communities to develop their Neighbourhood Plans. Although Neighbourhood Plans applied to urban and rural areas, people in the country were better placed to engage with them due to their well-established network of legally recognised parish councils (by contrast, in urban areas, there was a need to establish new neighbourhood forums). Neighbourhood Plans give local people some say in where development occurs, offer opportunities to suggest new forms of development but, at the same time, prevent communities from objecting to developments agreed in local plans (Parker et al., 2015; Gallent, 2013). However, they are usually written by 'well-meaning, well-educated people living in nice places – mostly rural – with time on their hands' (Peter Hall, 2011, quoted in Gallent et al., 2015). More cynically, they continue to reflect class interests.

In an effort to give community-based planning more teeth, partnership working has become a key aspect of rural development policy in many countries. Partnerships enrol private, voluntary and state agencies to '*empower themselves* by blending resources, skills and purposes with others. The capacity to get things done no longer lies (if it ever did) with government power or authority in one place' (Kearns and Paddison, 2000: 847). Edwards et al. (2001) identify three different types of partnerships:

1. strategic partnerships that coordinate policies and initiatives of different agencies in specific rural areas, such as the Devon Strategic Partnership;
2. delivery partnerships that provide services at a local level, such as housing partnerships or crime and safety partnerships;
3. consultative partnerships that engage local communities to gauge views and opinions, such as parish plans.

Box 10.3 Rural Citizenship

Citizenship refers to a person's relationship with a political unit, usually a nation state (Yarwood, 2014). With this association comes an obligation for the citizen to fulfil particular duties and the state to assure certain rights. These ideas can be applied in a rural context to examine the rights afforded to people living in the country and the kinds of duties or actions they undertake to assert them (Yarwood, 2017).

Marshall (1950) proposed that all citizens are entitled to civic, political and social rights, but it has been argued that citizens living in rural areas have fewer rights than those living in cities. Tonts and Larsen (2002: 135) note that 'as governments withdraw, or fail to provide, certain services and infrastructure the human rights of rural people are diminished'. Thus, a lack of policing impacts the civic right to justice, political rights are eroded in by decisions made in distant, urban places, and a closure of a doctor's surgery diminishes the social right to health provision. The language of rights is sometimes used to prompt better provision of rural services. In the UK, 'rural proofing' was introduced in 1999 as a mandatory part of the policy process and obliges policy-makers to consider the impacts of their policies on rural places.

Citizens can assert their political and civic *rights* by participating in formal decision-making and structures of governance. Thus, somebody concerned about the environmental impacts of a new road could attend planning enquiries and argue against it. Similarly, a citizen could petition elected representatives or stand for office themselves, perhaps in a parish council, to effect change at a local level.

Indeed, as this chapter has shown, there has been a greater expectation for rural citizens to take part in community-based initiatives. This form of voluntary participation is referred to as 'active' citizenship and emphasises *the duty* of citizens to contribute to their localities. There are three reasons why 'active citizenship' has been promoted in rural areas. First, rural areas have been more likely to suffer from the withdrawal of the state services and are, therefore, more likely to rely on citizen action to fill gaps in state provision. Second, there has been a long-standing obligation, evidenced in many countryside policies, that rural areas *should* provide their own needs. Examples include community-run shops, voluntary policing, locally built housing and health care. Finally, rural areas are better placed to engage in this form of local participation. The lowest tier of formal government, such as parish councils in England, are found in rural places. Rural residents therefore have more opportunities to engage with formal government and decision-making than their urban counterparts. Active citizenship has been criticised for promoting duties over rights, volunteering above political participation and, implicitly or otherwise, supporting the neoliberal roll-back of the state.

By contrast, it is possible to trace a range of 'deviant' actions that challenge state and corporatist power to assert social and political rights. In contrast to active citizenship, which is largely focused on changing neighbourhoods, *activist* citizenship is global in its concerns and reach (Parker, 1999). In the past, rural activists tended to come from *outside* rather than *within* rural places (Woods, 2011a), as evidenced by protests about road building, fracking or other environmental protests but there has been a growth of protests from and about rural areas (Woods, 2016). These have centred on three sets of rights: the decline of rural services and the social rights associated with them; the political right to demand state intervention to protect rural interests, especially the livelihoods of farmers; and, conversely, the social, cultural and civil rights to follow particular lifestyles, such as hunting, without regulation from the state. Groups such as the Countryside Alliance in the UK and Land Schafft Verbindung (Land Creates Connections) in German exemplify these movements.

These New Social Movements (NSM) are autonomous, pluralistic and transnational, occasionally crystallising in particular (and often urban) protest sites. Their actions represent a transnationalism that is concerned with global rather than national citizenship.

Government is thus seen to be one component of governance and, increasingly, having an enabling or coordinating role within new power structures. Many funding streams, such as LEADER (Box 10.1) or local housing (Chapter 11), require projects to be delivered through partnerships. In this way, some partnerships have been empowered to effect change in a rural place and, at the same time, involve a wider range of decision-makers. This has highlighted the importance of rural citizenship and the ways in which individuals can participate in decision-making or take action for change (Box 10.3).

At the same time, this leads to a confusing mix of decision-makers (Winter, 2006) who manage a nobody-in-charge world. Rather than offering a radically new form of governance, some commentators have suggested that partnership approaches represent a form of 'governance by distance' (Rose, 1996), whereby the state encourages, persuades, enables and supports community-based groups to implement government policy at a local level. While it is clear that new stakeholders have emerged in rural decision-making, it appears that power is largely held by elite groups, albeit different ones, and others remain excluded from decision-making.

10.4 Conclusions

There have been significant shifts in rural policy. These have aimed to support:

1. new agricultural regimes based on multifunctional practices;
2. the diversification of the rural economy beyond farming activities;
3. endogenous forms of development (Table 10.1);
4. multi-scalar actions;
5. partnerships and consensual decision-making;

6. territorial rather than sectorial development;
7. social, cultural, economic and environmental development.

(Van der Ploeg, 2000; Pemberton, 2019; Lapping, 2019)

There has also been a shift from a rural politics, which was concerned with farming and land use, towards a 'politics of the rural', which reflects that 'the meaning and regulation of rurality itself is the primary focus of conflict and debate' (Woods, 2006: 58). Despite changes in planning and policy, decision-making continues to reflect and support class interests. The following chapter uses the example of housing to examine this in more depth.

CHAPTER

11 Housing, Homelessness and Home

11.1 Introduction

Housing matters. It is more than simply a shelter (although many are in need of this basic provision): it is a home, an investment, somewhere that shapes identity and, importantly, the means by which people can live in the country. Yet in rural areas it can be difficult for many people, especially the young or those on low wages, to access housing and the benefits it brings.

Wealthier, middle-class incomers are often accused of outbidding local people for housing, leading to inflated house prices and forcing locals out of their communities. They have also been associated with the drawbridge effect, whereby wealthy incomers form a vociferous, well-organised and articulate anti-development lobby to object to further housing (Herington, 1984) or, put more irreverently, 'the last one in runs the village preservation society'!

The reality is more complex than this and reflects restrictive planning policies, high house prices and, to some extent, the incidence of second homes. This combination has made housing unaffordable for many people, forcing them to move away from their local communities. This chapter examines why housing remains out of reach for many.

11.2 Housing Tenure

Tenure is important because it determines who can access housing in particular places. Tenure is the legal status by which a property is occupied and there are three main categories: owner-occupied, privately rented and socially rented.[1] Owner-occupied or private accommodation is owned by the people who live in it (either permanently or for part of the year). Houses may be bought outright or, more usually, purchased with a mortgage (a long-term loan) that is based on a multiple of household income. Privately rented housing is owned by a private landlord and rented to tenants, usually at market rates. By contrast, social housing belongs to the government, housing association or a charity (all known as social landlords) and is rented to tenants at lower-than-market rents.

DOI: 10.4324/9780429448966-14

The balance between owner-occupied, socially rented and privately rented housing varies between countries. In the EU, for example, rates of owner occupancy range from 51.4% in Germany to 96.4% in Romania (Eurosat, 2020a), reflecting national differences in attitudes towards property ownership. In some countries, such as the UK, it is viewed as desirable to own property as it is regarded as an investment as well as home. This ideology was reflected by the sales of state-owned council houses to long-term tenants in the 1980s. In Sweden, there is greater reliance on state-owned housing, reflecting a political aim of social equality (Magnusson and Turner, 2008).

1. Owner-Occupied Housing

There is usually a greater reliance on private housing in rural places. In sparsely populated hamlets in England, 75% of homes were owner-occupied, compared to 67% in urban areas (Department of Environment Food and Rural Affairs, 2022). Similarly, in the USA, home ownership in rural and small towns is 72%, compared to a national average of 65% (Scally et al., 2018). In countries with high rates of owner-occupation, access to rural housing is therefore largely determined by the ability to *buy* a house. This is problematic because house prices are too expensive for many people to afford.

The cost of housing has been inflated by a very high demand for housing, coupled with a low supply of new homes. Demand has been driven by counter-urbanisation (Chapter 7), a desire for rural lifestyles (Gillon, 2014; Garrod et al., 2006) and international investment (Phillips and Smith, 2018a). Supply has been restricted by planning policies and opposition to house building (Gallent et al., 2015) that reflect national differences in the ways that town and country are perceived (Gallent et al., 2003b). In Sweden, late urbanisation, strong intra-regional family ties and a high degree of second home ownership (Gallent et al., 2003b) have led to a close connection between urban and rural places. As a result, urbanisation is not regarded as a threat and there is a more liberal approach to allowing houses to be built. Similarly, there has been little regulation of housing in the Republic of Ireland, which has led to 'ribbon development' (new housing being constructed along a major road) and unaesthetic buildings (Gkartzios and Shucksmith, 2015; Scott and Murray, 2009).

In the UK, the expansion of urban areas has been viewed as a threat to rural life, and consequently, there are stringent controls on construction (Gallent et al., 2008). House-building is generally prohibited in the open countryside (there are exceptions for agricultural dwellings), and if land does become available for housing, it is sold at a premium. Consequently, house prices are often significantly higher than local wages, making it difficult to apply for a mortgage or save enough to pay a deposit on a house (Gallent, 2019). In the Cotswolds, England, there is a property-price-to-wage ratio of 13.5 to 1, meaning that the average property costs 13.5 times more than the average annual wages (Office for National Statistics, 2019). Given this gap, it is particularly challenging for young people to save enough money for a deposit on a house. Following the financial crisis of 2008 (which was itself triggered by the 'sub-prime' market or banks granting mortgages to people who did not have the means to repay them), banks have become more stringent in lending money, making it harder for some households to obtain mortgages. This means that it is difficult, if not impossible, for some people to live in the places where they work, have family or have grown up. Gkartzios and Shucksmith (2015) have referred to this as a 'spatial apartheid' whereby wealth determines who can live in the countryside. A decline in affordable rented housing has further intensified these exclusionary processes.

2. Socially Rented Housing

Social housing is delivered by the government or housing associations at affordable rents. It is allocated using criteria that determine who is in most need of housing. Sometimes these can prioritise local people, and indeed, some social housing is built specifically to meet local needs (Gallent and Scott, 2017). It can, and should, facilitate access to affordable housing and allow people on lower incomes to live in the country.

In countries that value private ownership, rates of affordable housing are woefully low. In 2006, the Commission for Rural Communities estimated a need for 10,000–15,000 more affordable homes in England. Yet between 2017–2018, only 8,510 homes were built by social landlords, compared to 33,370 private homes (Office for National Statistics, 2019). In the UK, state investment in affordable, social housing has declined because of budgetary restraints, an ideology of home ownership and a lack of political will (Satsangi and Gallent, 2010).

Building new social housing is particularly difficult when planning policies emphasise the conservation of open land (Chapter 10). It is rare that land is made available for housing, and when it is, the high cost of land favours private developers, who can afford to pay more than social landlords. Efforts to build social housing can also be met with opposition from property owners keen to maintain a vision of rural exclusivity (Yarwood, 2002b). Furthermore, the sale of social housing has reduced the stock of properties at affordable rents, changed the character of some estates and made it difficult for future generations to find affordable accommodation (Chaney and Sherwood, 2000; Forrest and Murie, 1992).

3. Privately Rented Housing

Privately rented housing can make a substantial contribution to housing but has received relatively little attention from rural geographers (Satsangi, 2005). In some rural regions, considerable numbers of farms and housing may be owned by large landowners (Murdoch et al., 2003) that are rented to tenants (Figure 11.1). This may be linked to employment, such as a home rented to a farmworker by their employer or a farmhouse occupied by tenant farmers.

Elsewhere, competition from holiday accommodation can push the cost of renting to unaffordable levels or force tenants into short-term (often winter) leases. Of late, there has been concern Airbnb is proving more lucrative to landlords and reducing long-term rented stock. The quality of rented accommodation may also be poor. Trailer parks in the USA, for example, are often located on the margins of settlements with poor access to resources (MacTavish, 2020) and reputations for crime (Notter et al., 2008; MacTavish et al., 2006).

11.3 Local Housing Provision: The Case of the United Kingdom

The obvious solution to a lack of affordable housing is to build more housing and make this available to those in most need. Plaid Cmyru, the Welsh nationalist party, have consistently called for bans on the sales of second homes and some governments, including those in New Zealand and Denmark, have restricted the sale of property to citizens and residents of their countries. But given the importance of the private market, governments are reluctant to intervene directly in the sales of existing properties. Instead, they have used planning regulations and planning permissions to determine who can buy new houses. This is a subtle but important difference – while there is no legislation in the UK to limit the sale of *existing* stock, there are rules that limit who can buy *new* stock (Gallent et al., 2020).

Governments have attempted to regulate the building of new homes for local people, often using complicated planning policies (Shucksmith, 1991). One way of doing this is to compel developers to build a certain number of affordable homes when building new housing. The rates vary (for example, it was as high as 60% in the South Hams in Devon) but can be contested by developers who, instead, might choose to build in urban areas where they can build more private homes and make more profit.

Another route has been through 'exceptions' sites. Here planning permission is given to build on land where housing would otherwise be prohibited. To do this, certain criteria need to be fulfilled – namely, that there is a proven need for affordable housing and that this housing will be owned by a social landlord in perpetuity. This cunningly circumvents competition on the open market as private developers are still not allowed to build on this land. Thus, it can be sold cheaply for social housing. The exceptions route has led to pockets of housing being built in some rural places, but it is a cumbersome and tenuous route that is liable to break down at many points in the development process (Yarwood, 2002b). It requires a suitable site for housing to be found, as well as a landowner willing to sell their land. Development will inevitably occur on green-field sites that are likely to attract opposition, particularly if the scheme is overlooked by existing private housing. Even if all these barriers are crossed, a lack of funding may prevent housing being built (in recent years, some private developments have been allowed on 'exceptions sites' to offset costs and incentivise development). For all these reasons, schemes can take many years to complete.

Over recent years, there has been a growth in Community Land Trusts (CLTs). These are locally run, 'democratic, non profit organisations that own and develop land for the benefit of the community. They typically provide affordable homes, community gardens, civic buildings, pubs, shops, shared workspace, energy schemes and conservation landscape' (Community Land Trust 2022). CLTs buy land and then work with a housing association to build affordable housing for local people. One example of a successful scheme is in the village of Christow, located in Dartmoor National Park, Devon (Figure 11.1). Christow Community Land Trust was established in 2011 by eight individuals in response to a housing needs survey that identified a lack of housing for young families and older people. Working in partnership with a local housing association and housing enabling officers, they purchased land from the parish council for a token amount and went on to build 18

FIGURE 11.1 Affordable housing built by Christow Community Land Trust, Devon.

houses, 14 of which were available for rent and 4 for sale at a discounted rate. The scheme was completed in 2016, with homes occupied by residents in the village. The houses were heated using passive technology, which reduced energy bills and their carbon footprint. Christow's scheme was driven by a small group of individuals who successfully aligned other local organisations, such as the parish council, to garner support for the scheme. CLTs can therefore be effective in providing for specific local needs, perhaps replicating the original function of housing associations.

As this example shows, CLTs can provide local housing but given their small scale, are unlikely to make significant impacts on housing shortages. Efforts to build on the open countryside are often met with fierce resistance that can delay, modify or, very often, end proposed developments, especially when the properties or those who will live there are seen as undesirable (Yarwood, 2002b; Scally and Tighe, 2015). Although objections are couched in terms of planning regulations or a wish to conserve the countryside, these are usually driven by a desire to maintain property prices (Saunders, 1984) or exclusive lifestyles and so reflect a form of NIMBYism (not in my backyard) (Matthews et al., 2015) (Table 11.1). The location of affordable housing often reflects local political support or opposition rather than where there is most need (Yarwood, 2002b).

Specifically, Mark Shucksmith (1990) demonstrates that property disputes reflect the relative power and vested interests of three classes of people:

- Suppliers: builders and landowners. Builders will look to maximise profits when building houses by, for example, building luxury rather than starter homes that will sell at the highest profit. Landowners will seek to maximise profits on land sold for housing but, equally, may be reluctant to sell land if it remains profitable in farming. If landowners have a large number of tenancies, they may oppose other forms of rented housing in order to maintain the rental value of their properties (Satsangi, 2007).

- Owner-occupiers: have a vested interest in maintaining the value of their properties. They will, therefore, oppose new developments that reduce the value of their house – for example, those that block a view. They will buy and sell property to maximise financial returns.
- Non-owners: including those who rent from councils, social or private landlords, as well as the homeless. They would gain most from an increase in housebuilding, especially rented accommodation, and cheap rents. Some longer-established social tenants favour opportunities to buy their house.

TABLE 11.1 Property Classes

Property Class	
Low Income, Low Wealth	
Young households	Often unable to buy and denied social housing. Seek private rented winter lets, mobile homes or share with parents.
Tenants of a private rented or tied accommodation	Trapped in inadequate housing and with little prospect of council housing.
Pensioners	Retired from local employment. May need support accessing services. May find it hard to maintain home.
Tenants of a local authority (council) or housing association	A small group of people but relatively fortunate. May have had to move to a property. Opportunities to buy properties at discounted rates, further reducing socially rented stock.
More Prosperous Groups	
Owner-occupiers, farmers and landowners	Have the incentive to prevent development to keep a low-wage economy and maintain property prices.
Retirement or lifestyle migrants	Ready capital from the sale of a previous house; wide choice but will oppose development to maintain values and lifestyle. Not tied to a locality.
Holiday homes	May have less money to buy (given they have a main home to maintain) and will therefore compete with low-wealth groups for cheaper properties.
Commuters	Choose to live in a rural environment rather than a particular community but also have the desire to prevent development from maintaining property values.

Source: Shucksmith (1990)

It should be noted that the formation and operation of these property classes specifically reflect the UK, its property markers and planning policies. Indeed, much of the literature surrounding housing conflicts seems to have been written from a UK perspective. As such, it provides an example of how restrictive planning regimes and neoliberal housing policies combine to prevent the provision of housing to those most in need.

11.4 Second Homes: Cause or Symptom of Housing Shortage?

During the Covid-19 pandemic, many countries saw a flight of second home owners from cities to their rural residences (Gallent, 2020). This migration was met with hostility in many places, with second homeowners told to return to their main residence lest they bring the disease with them or overwhelm local medical services. In response, second home owners argued that they had their right to reside in a second home and had paid to access local services through taxes. These episodes reflect some tensions about second homes.

Second homes are a significant presence in many rural localities, although there is a geography to them. Figure 11.2 shows second home ownership in England and Wales in 2011, demonstrating significant clusters in coastal locations and national parks (Gallent et al., 2003a). One example is Salcombe in Devon, where it has been estimated that over half the houses in the centre of town are second homes. By

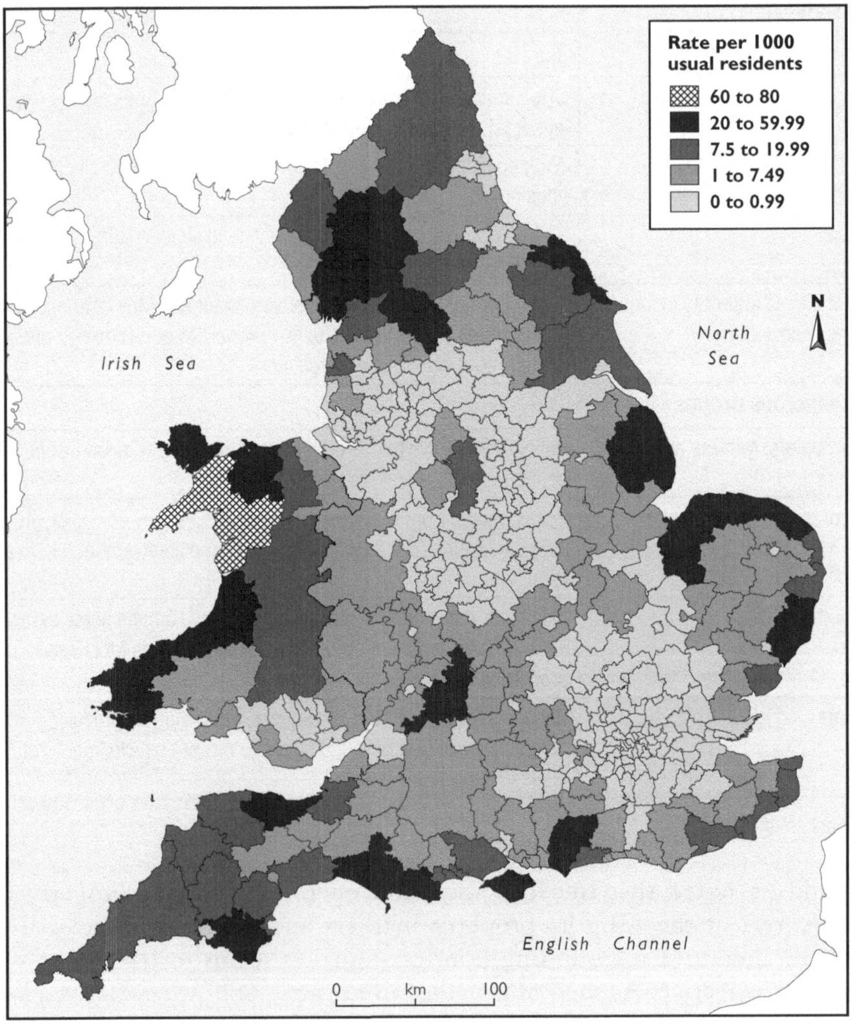

FIGURE 11.2 Second home ownership in England and Wales 2011.
Source: Office of National Statistics, 2012

contrast, many rural places, such as the Welsh Marches or central and northern England, have a very low number of houses. The impact and significance of second homes is therefore limited to particular localities.

Research by (Gallent et al., 2003a) Gallent et al. (2003a) has suggested that second homes may be a symptom rather than a cause of a decline in rural areas. Drawing on economic and housing data, they argue that second homes are often located in areas that are already in decline, offering opportunities for 'second home buyers [to] unwittingly exploit the weaknesses of the rural economy: they often find "bargain" properties in picturesque villages, but rarely question why property commands such a low price in the local market' (p. 280). Although second homes can be vilified by local residents, they are compounding existing affordability problems. Rather than seeking to block the purchase of second homes, Gallent et al. (2003a) advocate building more affordable rented accommodation.

In other places, second homes can be less contentious and are considered an important way to engage with nature (Rye and Gunnerud Berg, 2011). In Scandinavia, for example, summer houses support a tradition of spending some or all of the summer in a remote or coastal location (Müller, 2013). In Norway, there are 415,000 second homes (20% of local housing stock), and 40% of households either own or have access to one or several second home(s) (Overvåg and Berg, 2011). In Finland, second homes are located in remote, forest locations and often have basic facilities. They are associated with wilderness landscapes, traditional ideas of rustic life and as places to engage with nature through outdoor activities, such as wild swimming, saunas, hiking or foraging (Vepsäläinen and Pitkänen, 2010).

In South Africa, Hoogendoorn and Visser (2015) contend that second homes create employment, contribute to tourism and are more widely accepted part of local communities. Gallent (2014) has also argued that second homes have a clear social value, drawing new populations into rural communities and 'increasing the connectivity of communities to new skills and knowledge' (188). Drawing on Heidegger's idea of dwelling, Gallent (2007) suggests that a connection with place and community predates obtaining a second home. For example, the decision to buy a home might reflect a personal association with a place that has been strengthened by successive holiday visits. The decision to buy and live in a second home also reflects that communities are increasingly mobile. Halfacree (2012) considers second homes a form of 'heterolocalism', recognising that identities and lifestyle are forged 'through multiple places that do not depend on the core sedentarist assumption of a single, settled home place' (214). Who is or is not considered 'local' has plural meanings and is, therefore, contested. Despite lengthy absences, second home owners may consider themselves local, although this view may not be shared by those who live in a place all year round.

These nuances may provide cold comfort for someone facing continuing difficulty accessing a place to live. While second homes are undoubtedly complex, they will continue to remain contentious in places where they are concentrated.

11.5 Rural Homelessness

Homelessness is often associated with urban areas because activities associated with it, such as rough sleeping or begging, are more visible in cities (Lawrence, 1995).

Rates of homelessness are also higher in the city than in the country (Milbourne, 2006). In January 2018, 51% of all homeless people in the USA lived in major cities, with 18% in 'largely rural' areas (Henry et al., 2018). Yet rates of rough or unsheltered living were higher in rural places (20.8% compared to 18%), reflecting a lack of services for homeless people in rural places.

To an extent, homelessness reflects the lack of affordable housing. Thus, young people may 'sofa-surf' (living temporarily in the spare space of someone's home) or live in family homes until a late age and may not see themselves as homeless (Cloke et al., 2002). Yet Cloke et al. (2002: 66) argue that there has been a 'discursive non-coupling of "rurality" and "homelessness"' so that homelessness is rarely associated with dominant, idyllic ideas of rurality. Woods, trailer parks, camping grounds, hostels and sofas in friends' houses continue to hide homeless people in the countryside (Baumohl, 1996; Cloke, 2002). To illustrate this, Cloke et al. (2000b: 87) draw on the example of a homeless military veteran who wore camouflage clothing and drew on fieldcraft skills to remain hidden:

> *I'd walk round like 10, 15 miles a day in the countryside, just carry on walking, walk across main roads, dive through hedges and you know, just doing things like that. Um, if I seen a rabbit I'd shoot it and um, find somewhere that was nice and isolated and then start cooking it. And then once I'd eaten it, stay around for 5 to 10 minutes and then, um, destroy the fire, make sure it looked like no-one had had a fire and then disappear.*

Homelessness also reflects other, often hidden, social issues. Drug and alcohol dependency, loss of employment, physical and mental abuse, mental illness and relationship breakdowns can lead to homelessness (Milbourne, 2006: 908). Domestic violence or family breakdown may force some people, particularly women, children and indigenous people, to flee home (Fitchen, 1992; Cummins et al., 1998). Waegemakers Schiff et al. (2016) also note that migrant workers and immigrants are more likely to live in poor or inadequate housing. While these issues also affect people living in urban areas, there are fewer services to support people with these needs in rural places (Coy et al., 2009). Consequently, homelessness continues to be ignored or downplayed by policy-makers and the public.

11.6 The Home

A home is more than simply a building. We have strong emotional attachments to our homes; they are places where we feel safe (or should feel safe) (Valentine, 1997b), allow us to express our individuality and reflect and shape social values. Although geographers have traditionally been concerned with public places, the geographies of the home are also significant (Blunt, 2006; Brickell, 2012). Blunt and Dowling (2006) suggest that these reflect:

1. The 'material and imaginative' space of the home – home is a site of consumption that reproduces ideas of rurality, small-town life and nature. Properties are renovated, decorated, changed, extended and decorated in 'tournaments of taste' (Cloke et al., 1995b) that assert class difference, reflect consumer trends and improve market values (Hoggart and Buller, 1995; Phillips, 2002). Rurality

is commodified, packaged and sold for consumption within the home through rustic furnishings and decorations, as well as activities such as baking or crafting, which represent how rurality itself has been valued (Hughes, 1997; McCarthy, 2006b; Yarwood and Shaw, 2010). The designer Laura Ashley, for example, specialised in producing domestic goods that reproduced the rural idyll were given and gained prominence in many homes (Pratt, 1996) and, in doing so, acted as a nostalgic counterpoint to the breakdown of nuclear families and the complexities of (post)modern living (Leslie, 1993; Woods et al., 2021).

Home is also an affective and emotional space 'of belonging and alienation, intimacy and violence, desire and fear, the home is invested with meanings, emotions, experiences and relationships that lie at the heart of human life' (Blunt and Varley, 2004: 3). It is associated with growing up, family relationships and the end-of-life (Hockey et al., 2005; Blunt, 2006). At its worst, it is a place of fear caused by domestic violence that is given rein by the privacy home affords (Little, 2017). At its best, it is a place of care (Lawson, 2007; Conradson, 2003). Kelly and Yarwood (2018) show how a sense of familiarity, combined with care by family members, provides a feeling of security for people living with dementia:

> *We used to put mum in the sheep shed . . . she used to love lambs . . . the sights, the sounds. We think she appreciated it . . . But yeah, that was her thing, and we couldn't have done that if she was in a [care] home.*
>
> *(p. 102)*

2. Home, power and identity – critical readings of the home point to the ways that it reproduces and enforces inequalities of age, gender, sexuality, ethnicity and class. Davidoff (1995: 42) considered that the 'very core of the [domestic] ideal was a home in a rural community'. Home, he continued, is sheltered and separate from public life and scrutiny, offering a setting to establish gendered, patriarchal relations. These, for example, are reflected in domestic skills, such as baking and crafting (Hughes, 1997). Kallis et al. (2020) demonstrate how ideals of Cypriot rural life have been applied to the homes of migrants in England. In particular, women have been expected to adhere to traditional sexual values, learn Cypriot cooking and maintain annual traditions. They have also used the space of their gardens to cultivate plants to make food associated with Cypriot villages.
3. Home as a 'multi-scalar' place – 'bound up with, rather than separate from wider power relations' (Blunt, 2006). Thus, the farm*house* provides a space both for living and conducting business (Whatmore, 1991b). Its domestic space maintains gender roles (see Chapter 17) that, in turn, reproduce the social structures and power relations that are relied upon to support the economic activities associated with family farming (Price and Evans, 2009). Consequently, public and private space becomes blurred.

The hybrid nature of home space is also reflected in the geographies of children in the country (Ward, 1988). Children are disciplined to behave in public spaces in ways that reflect the values of the private space of the home (Valentine, 1996). Home, therefore, imposes and installs a moral geography that is mapped out in public spaces (Panelli et al., 2002b; Leyshon, 2008a) and reinforces ideas about what behaviours are or are not associated with rural places (Giddings and Yarwood, 2005). At the same time, children have their own agency that can challenge these norms. Thus, behaviours and ideas learnt in public spaces, such as schools or online, may challenge those of the home (Staeheli, 2011).

11.7 Conclusions

A focus on housing and the home provides a window (no pun intended) into the processes that shape public and private spaces in rural places. It not only reveals the social structures that reproduce inequality but also underlines the significance of housing and the home in reproducing these. Although these are often couched in terms of wealthy newcomers and poor locals, this chapter has demonstrated that access to housing is a constituent of class and opportunity. In turn, housing affects how people can move or stay in the country, processes that are explored more fully in the idea of mobility.

Note

1 In some cases 'shared ownership' combines these tenures. For example, some houses are part-owned and part-rented from a housing association. This provides opportunities for people to buy a house by reducing the cost associated with a deposit and mortgage.

CHAPTER

12

Mobility and Immobility

12.1 Introduction

Rural places are not static (Massey, 1991). Rather, they are continually shaped and reshaped by incessant movements of people, goods and media (Adey, 2017; Sheller and Urry, 2006). Place, as Massey (2005: 141) argues, is a 'constellation of processes rather than a thing'. In her seminal work, *For Space*, Doreen Massey (2005) chooses a rural locality – Keswick in the Lake District – to illustrate this. She argues Keswick:

> *is the event of place. It is not just that old industries will die, that new ones may take their place. Not just that Hill farmers round here may one day abandon their long struggle, nor that that lovely old greengrocers is now all turned into a boutique selling tourist bric-a-brac. Nor, evidently, that my sister and I and a hundred other tourists soon must leave. It is also that the hills are rising, the landscape is being eroded and deposited; the climate is shifting; the very rocks themselves continue to move on. The elements of this 'place' will be, at different times and speeds, again dispersed.*
>
> *(Massey, 2005: 140-141)*

As Massey points out, tourists move in and out of Keswick, industries come and go, and even the hills themselves are moving, albeit slowly! By considering rural places as constantly changing, it becomes difficult to put boundaries around them. In this sense, *place* might be better thought of as verb rather than a noun – a movement rather than a location.

In rural geography, attention has mainly been given to long distance movement (Chapter 8), but as Paul Milbourne (2007) shows, a plethora of movements shape rural places:

> *Rural population change is . . . composed of movements into, out of, within and through rural places; journeys of a few yards as well as those of many hundreds of miles; linear flows between particular locations and more complex spatial patterns of movement; stops of a few hours, days or weeks as well as many decades; journeys of necessity and choice; economic and lifestyle-based movements; hyper and im-mobilities; conflicts and complementarities; uneven power relations*

DOI: 10.4324/9780429448966-15

and processes of marginalisation. It is these different mobilities, present in different combinations in different places, that produce the complexities of rural population change.

(385–386)

Tim Cresswell (2010) suggests that places are shaped by a 'politics of mobility' that draws on three elements:

1. A fact of physical movement – the physical ways in which ideas, people and goods move. For example, how many rural bus journeys are made in a year, or how well connected are places? This has been the traditional focus of rural transport geography.
2. A representation of movement – how movement is portrayed and represented by different media. These can be contradictory. For example, a rural bus journey might be advertised as convenient, yet social media reports may highlight late or cancelled journeys.
3. The experienced and embodied practice of movement – what is it like to experience travelling, both physically and emotionally? The traveller on the bus journey might experience delight at a stunning view or discomfort at having to wait for hours at a cold, isolated bus stop for the bus to arrive. How will this affect their relationship with rural places?

The following section illustrates these ideas by examining the mobilities of migrant agricultural workers in the EU.

12.2 Migrant Workers

The facts of physical movement. Woods (2007) notes that a significant feature of the global countryside is the migration of agricultural workers, often from rural places in one part of the world to rural places in another. Examples include migrants from Latin America to the USA, North Africa to Italy and Spain and the Pacific Islands to New Zealand.

In 1992, the Maastricht Treaty granted citizens of EU countries the right to move between other EU states and to live and work in them (Ferbrache and Yarwood, 2015). In 2004 the EU enlarged to encompass Poland, Estonia, Latvia, Lithuania, Hungary, Czech Republic, Slovakia, Slovenia, Cyprus and Malta, followed by Romania and Bulgaria in 2008. These accession states, as they are known, had considerably lower GDPs than existing EU states. This disparity, together with the right to work and movement as EU citizens, prompted migration to take up work in a range of jobs that offered relatively high wages, including manual farm work associated with the harvesting and processing of food.

As a consequence, it is estimated that migrants account for 5.5% of the EU's rural population (Natale et al., 2019). Between 2011 and 2017, the number of migrants in the agricultural workforce increased from 4.3% to 12.5%, with the highest increases in Spain, Italy and Denmark (Natale et al., 2019). Interestingly, this increase in the availability of labour led to a growth in certain crops, such as soft fruit and asparagus, that are labour-intensive (Figure 12.1)

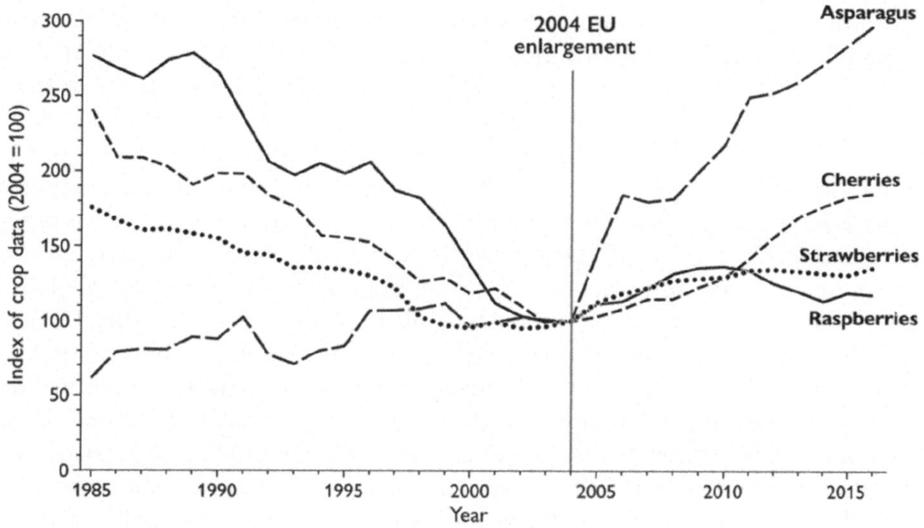

FIGURE 12.1 The impact of migrant labour on the production of selected crops in the UK.

One region to benefit from migrant labour was Herefordshire in the UK, where fruit production is an important farming activity (Storey, 2013). Workers from Poland were particularly important, accounting for 12% of the agricultural workforce (Table 12.1). These were mainly young men between 18 and 34 who would move to other low-paid, low-skilled jobs at the end of the harvest season. Most wished to work for five years before returning home.

TABLE 12.1 Migrant Workers in Hereford, UK, 2002–2007

Country of Origin	Numbers
Poland	4,400
Lithuania	938
Slovakia	720
Bulgaria	680
Latvia	400
Portugal	330
Hungary	270
Romania	210
India	200
South Africa	170
Philippines	160
Czech Republic	150
Estonia	100
Ukraine	100

Source: Dawney (2007); Storey (2013)

A representation of movement. While these data reveal some of the 'facts' (Cresswell, 2010) of economic migration, the way that this movement was *represented* revealed a series of tensions. Storey (2013) notes that the British tabloid *The Daily Mail* reported that 'Hereford has become a magnet for migrants (who) have swarmed to the area to work in the hospitality and catering trades and to pick or process fruit and vegetables' (*Daily Mail*, 14 March 2009). The term 'swarm' dehumanises migrants as it is more often used to describe locusts or insects. Elsewhere, the same newspaper reported that 'people have complained about eastern European migrant workers walking around Leominster, Herefordshire' (*Daily Mail*, 18 July 2012). Walking around a town is not a criminal activity! Instead, it reveals how migrant workers are regarded as 'out of place', something that was reinforced by graffiti on a town sign (Figure 12.1).

The experienced and embodied practice of movement. In response to hostility in Hereford, local authorities and agencies have made efforts to HIGHLIGHT the contribution of migrant workers to the rural economy as well as the social and cultural diversity of the countryside (Storey, 2013). In another study, Maher and Cawley (2016) show that Brazilian workers in Gort in the Republic of Ireland led to new shops, churches and festivals becoming incorporated into the area's cultural life. MacKrell and Pemberton (2018) also revealed that many migrant workers to the North West of England shared the idyllic vision of rural places and coupled them with their own national cultures of self-sufficiency.

Mobility is not simply about the movements of people but also how flows of cultural ideas and material goods connect places. Thus, Gort and Brazil, Hereford and Poland, and Norway and Romania are not only linked by the movement of people but also through cultural practices, flows of currency (as workers send money home) and communication. The result produces a form of 'translocal' space that is at once both locally distinctive but globally connected (Brickell and Datta, 2011). More specifically, 'transrurality' takes

> *account of the specificities of place and, at the same time, [pays] closer attention to the ways in which rurality is implicated in and implicates other spaces and places, not only with regard to its binary 'the urban' but also networks of spaces and places across different scales.*
>
> *Askins (2009: 379)*

Not all of these international networks are licit and migrant workers are more likely to be victims of global criminal activity. The International Labour Organization (ILO) estimates that nearly 25 million people are in forced labour, and of these, 11% work in agriculture or fishing (International Labour Office, 2017). This includes forms of bonded labour (when people become indentured due to debt), child slavery, forced marriage, descent slavery (being born into slavery) and human trafficking (Byrne and Smith, 2016). Transnational trafficking makes illicit links between the Global North and Global South, reflecting that rural areas are both the suppliers and receivers of immigrant labour (Woods, 2007). Although migrant labour is regulated by the state (Rye and Slettebak, 2020), illicit people-trafficking exploits vulnerable and powerless groups by paying no or low wages, stealing passports, and deducting money for substandard accommodation (Allamby et al., 2011). This is compounded because workers may be unaware of their rights (Rye and Slettebak, 2020), afraid of authority or isolated by language, rendering them powerless to complain (Allamby et al., 2011). Notoriously, in 2004 at least 23 Chinese migrants drowned in Morecombe Bay, UK, as they worked as cocklepickers for 11 pence an hour.

12.3 Everyday Mobilities

Although global mobilities are important, local movements also play a significant role in the constitution of rural localities. Drawing on a series of travel diaries (Box 12.1), Milbourne and Kitchen (2014) illustrate how everyday mobilities shape and reflect people's lives. Dorothy (Box 12.1), for example, is a pensioner with no car or internet, limiting her mobility. By contrast, Chris is also retired but, being wealthier, is able to travel when he needs to (for example, to the supermarket to shop) but chooses to walk frequently for leisure and exercise. Penny, also retired, cannot drive but has access to the internet, using it to pay bills, order home deliveries and keep in touch with relatives in Australia. Community buses and lifts from friends allow her to maintain contact outside the village. Jenny's ill health means frequent car trips to a hospital and make local walks problematic. Milbourne and Kitchen (2014) explain that these examples reveal some of the complexities of everyday mobility, as well as a reliance on private cars or community transport. Elsewhere, the paper also points to the significance of the internet, which allows services to be accessed and friends to be contacted, yet, at the same time, depends on reliable connections and broadband speeds.

Box 12.1 Everyday Travel Diaries of Residents in Rural Wales

Chris

- 24th February: Take car to Merthyr Tydfil to diagnose oil leak, two hour return trip. Myra walked the dog for me and then went for four hour ride on horse.
- 25th February: Walked to town and back to collect papers – total 3 miles, accompanied by Myra and the dog. Walked back in to town – friend had trouble with front door lock. Myra out on horse – 4 h.
- 26th February: Walked dog approx 2 miles. Short walk today. Then drove 20 miles to Llandrindod Wells for special ingredients for dinner party.
- 27th February: Get dressed for hill walking. Pack dog in car and drive to station (1 mile). Meet two friends and 4 dogs! Take train approx. 8 miles south then walk back to Llanwrtyd Wells via cross-country roads and hills. Picnic lunch, swift half at the local hotel, then home.
- 28th February: Walked dogs; we are looking after friends dog for two days. Rest of day spent in house.
- 1st March: Walked dog 3 miles. Called in on elderly neighbour to set up her new TV. Myra out on horse (3½ h). Walk 3 miles to pub – swift couple of halves. Home for dinner.
- 2nd March: Walked in to town after breakfast with dog. Fetched paper, sorted out world and agricultural problems with local farmers. Walked back home (6 miles total).

Dorothy

- 23rd February: No journey undertaken
- 24th February: No journey undertaken
- 25th February: No journey undertaken
- 26th February: No journey undertaken
- 27th February: Weekly journey to Builth Wells for shopping and to collect repeat prescription. Also to return and then choose library books from Builth library. Then to Builth Cooperative store which is the only supermarket for 17 miles. Transport courtesy of Llanwrtyd Wells and District Community Transport. Collected at my door 11.10 to 11.15 and returned at approx 13.50. This service is an absolute Godsend to those of who have no transport of our own!
- 28th February: No journey undertaken
- 1st March: No journey undertaken

Penelope

- 25th February: My husband John and I were picked up at the door by the Community Bus at 9.45am bound for Carmarthen. We get quite excited when we see all the displays in stores such as Marks and Spencer's, Wilkinson's and The Works, and spend more money than we should! Laden with bags, we rejoin the Community Bus at 2.00pm and head for home. We see the finest little lambs in the fields, catkins on the trees. We are deposited at our door at 4.00pm. It has been a successful day's outing and we are grateful for this service.
- 26th February: We are a bit tired today, after running around Carmarthen yesterday. This is not just physical – though this is part of it but living as we do in peaceful rural Wales, among the hills, fields and rivers woken by the dawn chorus. And hearing the breeze in the trees a day in town or city with all the noise and bustle, busy crowds of strangers hurrying, scurrying – with the traffic and plethora of sights all attracting attention, and decisions, can all feel quite draining (even for people born and bred to big city living)!
- 27th February: This morning. The butcher delivers the meat to the house (he had phoned as usual for the order on Monday afternoon). He also delivers bags of potatoes and other vegetables if we want them. John joined the Community Bus, bound for Builth Wells, where he would purchase fresh bread and greens for the weekend. He also bought a mobile phone as our old one – years old had ceased working! At 5.15pm friends collected me by car and we went to our weekly art class.
- 28th February: John went as usual to the village inn with our neighbour, Tony. Peter calls. He has brought coal and bird nuts for me.

- 1st March: We spend a quiet day reading the Sunday papers.
- 2nd March: I submitted meter readings to Eon online prior to next bill. I also sent emails to family members in Australia.
- 3rd March: George, the log man, arrived during the morning and he and John uploaded and stacked them. The young woman from the village comes to the door with fresh eggs. John went with our neighbour, Bob, in his car to the nearby village of Beulah for coal and wood strips and cream paint from the builders' merchants for the bathroom.

Jenny

- 25th February: Car journey to Carmarthen (20 miles each way) to collect bulk order of detergent from Morrisons [supermarket]. Car to local rubbish dump. Then on to Crymych to butcher (6 miles). Drove domestic help home (2 miles).
- 26th February: Car to Newcastle Emlyn (6 miles) for building society and prescriptions from chemist.
- 27th February: Car journey to Newcastle Emlyn (6 miles) for further prescriptions following visit by doctor.
- 28th February: Car to local shop for newspaper and small groceries (1.5 miles). This used to be the post office but that was closed last year. The range of groceries is minimal and highly priced, so we rarely make use of it. Bread can only be bought once a week.
- 1st March: Car to another local shop for Sunday newspaper (3 miles). This is the nearest post office. Short car trip to nearest post box. Only a quarter of a mile but involves a very steep hill, beyond the capacity of my husband or myself with our health problems.
- 2nd March: Car to Tesco at Cardigan (10 miles each way). Take home domestic help.
- 3rd March: Emergency journey by car to Carmarthen following visit by doctor. Carmarthen is 20 miles away but is our nearest hospital for any but the slightest treatment. The doctor was unwilling to involve ambulances or hospital cars, so we were forced to use our own transport.

Source: Milbourne and Kitchen (2014: 332–333)

Mobility is a fundamental aspect of rural citizenship (Cresswell, 2009). The ability to move allows people to access services – such as shops, schools, places of worship, voting booths and so on – that are central to their civic, social and political rights. For years, rural geographers have stressed the significance of transport to the country, noting the social and economic impacts of poor transport provision on rural places and people (Moseley, 1979; Higgs and White, 1997). Shaw (1979) coined the idea of 'mobility deprivation' to stress the significance of transport to rural people to gain access to sparsely located jobs, services and facilities. Isolation remains a significant issue for many rural people where physical and electronic infrastructure is poor (Park, 2017; Shergold and Parkhurst, 2012) and disproportionally affects the old, the young, women and the disabled (Gray et al., 2006). Immobility, as Cresswell (2010) reminds us, contributes to a politics of immobility: thus, a lack of transport dis-empowers some rural citizens and renders them unable to participate fully in society.

12.4 Countercultural Mobilities

The idea of mobility stresses that places are continually being reshaped and reconstituted by movement, yet despite this, rural life is often valued for being fixed and, apparently, unchanging. Hubbard (2005), for example, notes the social panic that accompanied plans to build an asylum centre on a former airbase in rural England. One resident stated:

> *This rural area is quite unsuitable for such a massive venture, housing mainly men who have no or little concept of our rural way of life ('DS' quoted in Hubbard (2010: 62).*

Travellers, including Romani, indigenous people, those living on canals or so-called New Age travellers, have also long been regarded with suspicion by some rural residents (Halfacree, 1996; Holloway, 2003). In a study of Appleby New Fair, one of Europe's biggest horse-trading events, Sarah Holloway (2004) examines how travellers are seen as both 'in and out of place' – at once seen as a traditional part of the countryside yet, at the same time, regarded as dirty and dangerous (Figure 12.2). Other studies have revealed how travelling people are associated with crime and distrust, leading to exclusionary policing practices (Yarwood and Gardner, 2000; Vanderbeck, 2003).

New Age Travellers (NATs) added to this moral panic in the 1990s. Ethnicity, appearance, cultural traditions and behaviours (Earle, 1994). Initially, many NATs moved between music festivals and peace camps before adopting a regular mobile lifestyle. Camping illegally, their numbers were swollen by people from urban areas

FIGURE 12.2 Travelling people in Worcestershire.

seeking an escape from economic hardship or, on occasion, a sense of freedom through illegal 'raves'. New Aged Travellers differed from more established travellers in terms of their ethnicity, appearance, cultural traditions and behaviours (Earle, 1994). Without irony, some politicians drew distinctions between new nomads and Gypsies, arguing the latter were important to the rural economy, despite centuries of hostility:

> *True gypsies have been with us for centuries. They have been tolerated – and indeed welcomed – in rural communities where they regularly assisted with the harvest . . . but today they have acquired their parasites, the hippies or drop outs – generically referred to as new age travellers – who do not work, who do not want to work, but who believe that because the gypsies have the apparent right to roam the countryside at will, they can do the same at the expense of the local taxpayer.*
>
> *(Member of House of Lords, quoted in Halfacree, 1996: 99)*

A growth in numbers, coupled with pressure from landowners, resulted in legislation that effectively criminalised their mobile lifestyles by taking away the right to reside in temporary sites (Hetherington, 2000; Halfacree, 1996). In France, up to 10,000 Roma people living in over 300 camps and squats in France were forcibly deported to their home states of Bulgaria and Romania, contravening EU law allowing freedom of movement and preventing the prosecution of people on the grounds of ethnicity (Nacu, 2011).

Historically, there have also been efforts to contain indigenous people that have often moved between places as part of a traditional lifestyle. In Australia, for example, a lack of permanent settlement meant that British colonisers declared *terra nullis*, or that the land was 'empty'. This not only ignored the millions of people already living in Australia but, at the time, meant that Australia could be settled (note, the term implies living in one place) according to law. This process led to indigenous Australians being denied their traditional, mobile lifestyles and being confined to particular places and settlements, often in isolated, poorly resourced areas (Yarwood, 2007b). This process has been repeated in many parts of the world, as evidenced, for example, in North American reservations. The denial of mobile lifestyles, amongst other things, has led to deprivation that is reflected in material poverty, poor health and issues with addiction.

These examples reflect an expectation that people should live static lives in rural places. By contrast, nomadic lifestyles have been regarded with suspicion and treated with repression. But this denies the significance of mobility, even to places that apparently seem fixed (Massey, 2005).

12.5 Conclusions

The idea of mobility has drawn attention to the ways that the countryside is shaped by movement (Peacock and Pemberton, 2019; Gieling et al., 2017). On the one hand, this reflects and continues a long history in rural geography of examining how (a lack of) transport impacts rural life (Dufhues et al., 2020; McMillan Lequieu, 2017; Burke and Jones, 2019). This tradition has generally focused on movement within rural places, highlighting problems associated with inaccessibility.

On the other hand, the mobility turn has also suggested that the countryside is far from isolated. Instead, it draws attention to the interdependence of the rural localities with other, often far away, places. It has contributed to a better understanding of the global countryside and the international flows and movements that are shaping it. As the next chapter shows, global perspectives are necessary to ensure a sustainable future for the countryside.

CHAPTER

13

Sustainability, Resilience and Rurality

13.1 Introduction

Climate change impacts, and is impacted by, the country. Changes in land use, especially through intensive farming, deforestation and urbanisation, have contributed significantly to global warming and have led to a decline in biodiversity. The countryside has also been affected by climate change, be it through catastrophic events, such as flooding or extreme weather, or longer-term changes to local climatic conditions. There is a need for sustainable development in order to ensure the long-term future of the countryside and enhance its resilience against future shocks (Woods, 2019; Austin et al., 2020; Essex et al., 2005). This chapter outlines some of the work that rural geographers have undertaken to examine the relationship between rurality and climate change (Argent, 2017b; Austin et al., 2020).

13.2 The Impact of Climate Change on Rural Areas

It is widely accepted that climate change is being driven by society. Many scholars accept that we are living in the Anthropocene, a new geological epoch that recognises people, rather than natural processes, are behind climate change (Castree, 2014; Crutzen, 2006). Many organisations, including national and local governments, have declared a 'climate emergency' to highlight the urgency of action needed to tackle climate change. The Intergovernmental Panel on Climate Change (IPCC) was established to provide scientific policy-makers with assessments on climate change and its implications and risk. In its last report, the IPCC (2021) concluded 'unequivocally' that human activities have contributed to an increased concentration of greenhouse gases (GHG) since 1750, which have 'likely' led to an increase in global temperatures of between 0.8°C to 1.3°C and 'very likely' to have contributed to the warming of the oceans, glacial retreat and rising sea levels. One consequence is that capitalise (CIDs) will have a greater impact on rural society and ecosystems (Figure 13.1). CIDs refer to extreme events, such as droughts or cold spells, as well longer-term changes to weather and climate.

DOI: 10.4324/9780429448966-16

Storms and flooding are the most significant extreme events in many regions (Table 13.1), and it is widely accepted that climate change will increase their severity and occurrence (Feyen et al., 2012; Thorne, 2014; Cloke et al., 2013). Arnell and Gosling (2016) predict that by 2020 there will be a 187% increase in global flood risk due to climate change, affecting 450 million people and 430,000 km^2 of cropland, especially in Africa, Asia, North America and Europe (IPPC, 2021). The impacts of flooding are usually local rather than national (Thorne, 2014) and in rural, coastal and mountainous areas flooding can have significant impacts on local agricultural production and settlements (Yeo, 2003; McEwen et al., 2002). One study in rural France revealed that flash floods cost in excess of €15,000 per inhabitant, largely due to damage to public infrastructure (Vinet, 2008). Flooding in rural areas augments social disadvantage by disrupting already fragile transport networks or livelihoods (Argent, 2017b). For example, in coastal areas of Nova Scotia, elderly people are particularly disadvantaged when flooding destroys or renders inaccessible services that support them (Krawchenko et al., 2016).

TABLE 13.1 Annual Impact of Extreme Events, 30-Year Average, 1990–2019

	Occurrences	**Deaths**	**People Affected (Millions)**	**Economic Losses (US Dollars)**
Drought	16	813	57.9	5.1
Earthquake	28	27,494	4.8	24.7
Epidemic	42	6,768	0.78	ND
Extreme Temperature	18	5,872	3.5	ND
Flood	138	6,672	102.7	25.7
Landslide	18	879	0.23	0.26
Storm	99	13,697	21.7	47.5
Volcanic activity	6	86	0.24	0.08
Wildfire	12	79	0.22	3.4

Source: Asian Disaster Reduction Center (2020)

Wildfires also impact significantly on rural places in many parts of the world (Table 13.1). In 2020, there were 7,606 fires in western USA (compared to 4,972 in 2019), which destroyed 6.7 million acres of land and caused $20 billion of damage (BBC, 2020). These included five of the largest fires in California's history. Fires are a natural, and indeed necessary, part of forest ecosystems, but until recently, their risks have been underestimated.

Drawing on evidence from South Australia, Bardsley et al. (2015:158) demonstrate that forests offer beauty and 'opportunities for solace, reflection, health, work and happiness' but many people are often oblivious to the risk of forest fires or, at best, view it as an acceptable trade-off for an improved lifestyle. Ironically, this has been exacerbated by successful attempts to manage fire risk. As fire management improved, people became more willing to live in areas that were at risk of wildfires. Climate change means that these dangers are being realised and that it is necessary to devise new ways of managing risk and raising awareness of wildfire (McGee and Russell, 2003; Bardsley et al., 2015). In some cases, indigenous histories are being drawn upon to re-educate people about fire hazards. There has also been an increasing reliance on rural fire services, many of which are staffed by volunteers (Table 13.2). Yet while the need for their services is growing, it is becoming harder to attract and retain enough volunteers due to increasing regulation of the voluntary sector and disinvestment from regional governments (O'Halloran and Davies, 2020). Given the climate and societal change, wildfires will continue to pose a significant risk in rural places.

TABLE 13.2 Voluntary Emergency Services

State/Territory	Fire Agencies	Number of Volunteers	Number of Rural Brigades
Australian Capital Territory	ACT Rural Fire Service	413	9
New South Wales	NSW Rural Fire Service	72,491	2,002
Northern Territories	Bushfires NT	500	22
Queensland	Queensland Rural Fire Service	33,000	1,400
South Australia	SA Country Fire Service	13,500	425
Tasmania	Tasmania Fire Service	4,800	230
Victoria	Country Fire Authority	54,621	1,179
Western Australia	Department of Fire and Emergency Services (Rural Fire Division), Local Government Bush Fire Brigades	25,000	750

Source: O'Halloran and Davies (2020)

Fires and floods are dramatic and have significant impacts on local places. However, longer-term changes in temperature, rainfall and winds are of greater significance at regional and global scales. There has been a northward shift in climatic zones, which has resulted in a longer growing season in the Northern Hemisphere and increased opportunities for production (IPPC, 2021). By contrast, climate change will have the greatest impact on countries that make the least contribution to it (Meyer, 2019): it has been projected that rising temperatures in the Southern Hemisphere will pose significant challenges to the production of food (IPPC, 2021). One IPPC report predicts:

> *Major future rural impacts are expected in the near term and beyond through impacts on water availability and supply, food security, and agricultural incomes, including shifts in production areas of food and non-food crops across the world (high confidence). These impacts are expected to disproportionately affect the welfare of the poor in rural areas, such as female-headed households and those with limited access to land, modern agricultural inputs, infrastructure, and education.*
>
> *(IPCC, 2014)*

Countries in the Global South are more likely to experience declines in harvest due to extreme weather events, particularly droughts. In Africa, this may contribute to a fall in agricultural output of between 15% and 30% this century (Food and Agriculture Organization of the United Nations (FAO), 2009). One impact of this is that countries in the Global South will become more reliant on imported food and, consequently, more vulnerable to rising prices and supply chain shocks (Mbow et al., 2019).

13.3 The Impact of Rural Areas on Climate Change

Activities in rural areas also make significant contributions to climate change and environmental degradation. Agriculture, forestry and other land uses (AFOLU) contribute nearly a quarter of global GHG emissions, mainly through deforestation, methane emissions from livestock and mismanagement of soil (Smith et al., 2014). Agriculture alone contributes at least 14% to global greenhouse gas (GHG) emissions (6.8 Gt of CO2), although this varies depending on the type and location of farming (FAO, 2021).

Although 'food miles' – the distance food is transported from field to plate – has gained considerable public attention, most emissions occur during the production of food rather than during its transport, storage and cooking (Oglethorpe, 2009). Based on the carbon footprint of food consumed by an average US household, Weber and Matthews (2008) calculate that 83% of carbon emissions come from the production of food, with only 11% from transportation and 4% from its retail. Indeed, some locally produced food may have a higher carbon footprint than imported goods (Coley et al., 2011).

Drawing on data from 38,700 farms and 1,600 food processors, Poore and Nemecek (2018) conclude that the food supply chain contributes to a quarter (26% or 13.7 billion tons) of anthropogenic GHG emissions, as well causing 32% of acidification and 78% of eutrophication. These data include emissions associated with off-farm activities, such as transport or manufacture, as well as emissions during planting, growth, harvesting, processing and storage. They also estimate that non-food agriculture, such as growing crops or biofuels or keeping sheep for wool, and deforestation, contribute a further 2.8 billion tons of GHGs.

TABLE 13.3 Water Consumed in the Production of Food

Typical Values for the Volume of Water Required to Produce Common Foodstuffs		
Foodstuff	Quantity	Water Consumption (in Litres)
Chocolate	1 kg	17,196
Beef	1 kg	15,415
Sheep Meat	1 kg	10,412
Pork	1 kg	5,988
Butter	1 kg	5,553
Chicken meat	1 kg	4,325
Cheese	1 kg	3,178
Olives	1 kg	3,025
Rice	1 kg	2,497
Cotton	1 @ 250 g	2,495

Pasta (dry)	1 kg	1,849
Bread	1 kg	1,608
Pizza	1 unit	1,239
Apple	1 kg	822
Banana	1 kg	790
Potatoes	1 kg	287
Milk	1 × 250 ml glass	255
Cabbage	1 kg	237
Tomato	1 kg	214
Egg	1	196
Wine	1 × 250 ml glass	109
Beer	1 × 250 ml glass	74
Tea	1 × 250 ml cup	27

Source: Institute of Mechanical Engineers (2013)
(BBC, 2019)

TABLE 13.4 Carbon Emissions for Different Food Types The data show GHG emissions over one year if one portion is eaten per day (for example, eating two small potatoes a day will result in 16 g of GHG emissions per year).

	GHG Emissions Over the Year per Portion, per Day (in Kilogrammes)
Apple	12
Avocado	72
Banana	25
Beans	36
Beef (75 g)	2,820
Beer (1 pint)	243
Berries or grapes (handful)	44
Bread (1 slice)	21
Cheese (30 g)	325
Chicken (75 g)	47
Chocolate, dark (1 bar)	541
Chocolate, milk (1 bar)	375
Citrus Fruit (1 serving)	11
Coffee (1 cup)	155
Eggs (2 pieces)	202
Fish (farmed)	140
Lamb (75 g)	1,582
Milk (almond) (200 ml)	51
Milk (dairy) (200 ml)	229
Milk (oat) (200 ml)	65
Milk (rice) (200 ml)	86

Milk (soy) (200 ml)	71
Nuts (handful)	5
Oatmeal (3 tablespoons)	38
Pasta (75 g)	43
Peas (80 g)	6
Pork (75 g)	656
Potatoes (2 pieces, small)	16
Prawns (farmed, 5 pieces, king-size)	1,256
Rice (3 tablespoons)	121
Tea (1 cup)	15
Tofu (100 g)	58
Tomatoes	60
Wine	114

Source: BBC (2019); Poore and Nemecek (2018)

Many commentators have therefore advocated a change in diet, rather than a reduction in food miles, to reduce the environmental impact of food production (Weber and Sultana, 2013; Sadhukhan et al., 2020). This is because meat and dairy production uses significantly more water (Table 13.3) and contributes substantially more GHGs (for example, through methane emissions) than plant-based food (Table 13.4). Yet between 1963 and 2013, there was a ninefold increase in global meat consumption and the average person doubling their intake from 23 kg to 43 kg per year (Weis, 2013).

Questions have also been raised about the sustainability of rural lifestyles in relation to urban ones (Champion, 2009). Although cities contribute around 75% of global CO2 emissions (Smith et al., 2014), higher population densities and multiple-occupancy buildings mean that urban areas often have lower carbon emissions *per capita* than rural places (Heinonen and Junnila, 2011). Rural residents, especially middle-class migrants, commute longer distances and are less likely to use public transport (but are more likely to work at home) (Brown et al., 2015; Andersson et al., 2018). Taylor (2008) has suggested that this, coupled with restrictive planning policies, can lead to a 'sustainability trap' in rural places (Box 13.1).

Box 13.1 The Sustainability Trap

Lord Taylor (2008) argues that rural localities are liable to fall into a sustainability 'trap', especially if sustainability is defined too narrowly. He contends that a desire to preserve the countryside and reduce travel, often under the banner of sustainability, has led to a concentration of development in larger rural settlements. One consequence is that villages that are not deemed suitable for development remain attractive for middle-class migrants as they conform to a particular, preserved vision of rurality. The consequence is that there are further demands for conservation and preservation on the grounds of sustainability. As younger families leave, schools and other services close, rendering village life even less viable for lower-income families. This can lead to reverse commuting where middle-class rural dwellers drive to urban areas while those employed in low-wage rural jobs commute from urban (or even internationally) to the country. Echoing earlier critiques of 'key settlement' policies (Cloke, 1979; Pattison, 2004), Taylor refers to this as a 'sustainability trap' that does little for the environment and disadvantages lower-income groups.

Taylor (2008) reiterates the need for wider definitions of sustainability that include social and economic, as well as environmental, criteria. He advocates the development of new 'garden villages' that are similar to, but are smaller than, new towns (Taylor and Walker, 2015). These would be based on new or existing urban settlements and designed to incorporate community services, transport, employment and appropriate infrastructure. Development would therefore remain very concentrated, but he argues, this would allow the building of affordable housing, the preservation of historic settlements and improvements in suburban design (Lord Taylor of Goss Moor et al., 2022).

Yet wealth and consumption, rather than urban or rural location, are the biggest determinants of carbon emissions. Ottelin et al. (2019) suggest that the carbon footprints of urban areas would be 7% lower than rural areas *if their income and household characteristics were equal*. However, as income and consumption rates are generally higher in urban areas, the carbon footprint of cities is usually greater than that of the country. Consequently, the environmental impact of urban and rural living varies both globally and regionally, depending on how rapidly a country is urbanising and its standard of living.

Rapid urbanisation increases a city's carbon footprint, but this can vary considerably depending on whether urbanisation or counterurbanisation is more dominant (Food and Agriculture Organization of the United Nations [FAO], 2009). Drawing on an analysis of carbon emissions, Ottelin et al. (2019) conclude that rural areas have a higher carbon footprint than cities in Western Europe. This is reversed in Eastern Europe and in other countries where there are large disparities in income (Figure 13.1). It is, therefore, important to ask where high-income consumers live and to understand the social and economic drivers of consumption rather than ascribing environmental impact to an urban or rural location.

It also underlines a need for sustainable development policies to address economic and social, as well as environmental, goals. At a global level, the Brundtland Commission famously stated 'Humanity has the ability to make development sustainable to ensure that it meets the needs of the present without compromising the ability of future generations to meet their own needs' (United Nations, 1987: 16). The United Nations Development Goals (Figure 13.2) recognise a need to achieve 17

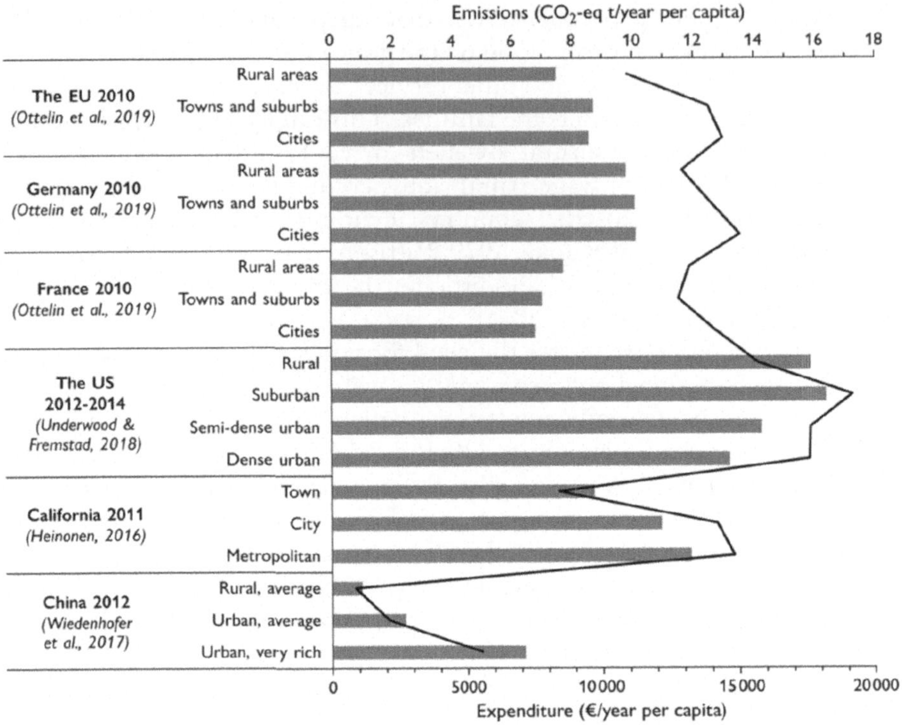

FIGURE 13.1 Carbon emissions and expenditure by urban and rural location.
Source: Ottelin et al. (2019)

different targets in order to achieve this. These ambitions have led to different eco-centric and technocentric approaches to try and realise these goals in rural places (Robinson, 2009).

Ecocentric approaches are linked to no- or low-growth scenarios. These emphasise eco-friendly practices, such as organic farming and a corresponding change in diets and consumption, including the adoption of plant-based diets. There is an emphasis on local food production and the reduction of transport. These initiatives are often bound into wider, sometimes radical initiatives to encourage sustainable living. One example is the Transition Town movement, which is seeking to shift places away from oil dependency towards a local, circular economy (Bailey et al., 2010).

Technocentric approaches have been more closely woven into agricultural policies that focus on modifying existing agricultural systems while maintaining high productivity (Figure 13.3). The Common Agricultural Policy (CAP), for example, is placing greater emphasis on actions to support environmental-beneficial farming and is seen by some as the new, 'third pillar' of CAP (in addition to economic and social objectives). Examples also include growing new crops, reducing livestock numbers, adopting environmentally friendly practices or embracing greener or smarter technologies (Holloway, 1999; Abid et al., 2016; Wang et al., 2021; Talanow et al., 2021). Perhaps the most radical technocentric approach has been the development of in vitro or artificial meat that is grown in a laboratory. This has the potential to significantly reduce the use of land, energy and water in the

FIGURE 13.2 The United Nations 17 Goals for Sustainable Development.
Source: United Nations (2022)

FIGURE 13.3 Solar farm on the edge of Dartmoor, England.

production of meat, albeit at significantly higher economic costs (Tuomisto and Teixeira de Mattos, 2011).

Drawing on the experiences of farmers in Sweden and Finland, Juhola et al. (2017) show that farmers respond to the challenges of sustainable development in three main ways:

1. Incremental: 'changes within the system, whereby most of its functions continue in their pre-existing trajectory' (p. 30). This might include adopting new breeds of livestock to help manage ecosystems or taking measures to reduce energy use.
2. Systems adaptation: 'changes to an existing farm system that changes some fundamental features of the system' (p. 30). This might include diversification or switching from meat to crop production or adopting new crops.
3. Transformational adoption: significant changes to a current system or moving to a new system, such as rewilding.

Based on their evidence, most change has been incremental or systematic with only some transformational change. Farmers also tended to be driven by economics rather than climate change *per se*, pointing to the need for robust agricultural policies to deliver sustainable farming.

Yet the practice of sustainable development has also been seen as something of a contradiction (Robinson, 2009), with farmers and policy-makers struggling to find a balance between competing social, economic and environmental demands. Sustainable development is most successful when it is supported, and develops, strong social capital (Box 13.2) (Mueller and Tickamyer, 2020; Juhola et al., 2017; Abid et al., 2016; Meyer, 2019).

Box 13.2 Windfarms

It is a trite observation that environmental problems, though they closely affect municipal laws, are essentially international and that the main structure of control can therefore be no other than that of international law.

(Sands and Peel, 2012: xxii)

Windfarms offer one way of generating renewable electricity and contributing to international treaties to reduce global warming (Figure 13.4). Although some windfarms are located off-shore, many are situated in rural areas that have consistent, high wind speeds (Naumann and Rudolph, 2020). Given these are in exposed, often 'wild' places, wind turbines are highly contentious and are frequently the subject of intense debate (Woods, 2003).

Many view windfarms as 'mechanical weeds', whose scale disrupts well-balanced, stable and, apparently, unchanging landscapes (Brittan, 2001). Objections to windfarms include their visual impact, ecological damage (for example, bird deaths), the sound of the blades, changes to land use, archaeological damage and electromagnetic intrusion. Others have questioned whether they bring any social or economic benefits to local communities (Munday et al., 2011: 4).

FIGURE 13.4 A windfarm near Fordsham, Cheshire, UK.
Source: Photo by Ruth Yarwood

In an analysis of windfarm developments in Wales, Woods (2003: 287) notes a politics of the rural, in which complex 'discourses of nature, landscape, environment and rurality' are deployed by supporters and opponents of windfarms. Many farmers, he notes, support windfarms in ways that reflect productivist values of the landscape. By contrast, opponents regard turbines as intrusions on 'natural' or 'unspoilt' landscapes and charge local people to defend unique and fragile environments. In some places, windfarms are novel tourist attractions but, elsewhere, might also be viewed as spoiling 'natural' landscapes sought by the tourist gaze (Strachan and Lal, 2004). This is reflected in a blog entry that refers to 'windfarmication', capturing a sense that windfarms are not only prevalent and destructive but are greedy intrusions from outside the area:

> *Welcome to the Highlands, the land of sparkling Burns, The heather hughed Glens, High Mountains where the Eagles soar, a land of Deer and Salmon, Kilts and Pipes. The Corbetts, the Grahams and the Monros. A place to revitalise the Spirit and the Soul. This is a land of proud people, people that will give any man the time of day. But today a certain sadness pervades all. In a desperate drive for fame our politicians have sold Scotland and its wild places to the lowest bidder. The march of the wind factories is heard in the Glens. Tourism for Scotland is dead. Our way of life crushed beneath the greed of mostly foreign adventurers and aided by our Government and Planners.*
>
> *(Quixote, 2011)*

Perceptions of windfarms vary over time and space: there are frequently high levels of approval prior to development, a decline during development, followed by increasing levels of approval as people become familiar with the development (Devine-Wright, 2005). However, given varied and nuanced responses by different individuals (Wilson and Dyke, 2016), it is difficult to generalise about responses to windfarms. Rather, community networks, advocacy campaigns, a sense of place and local politics influence how windfarms and other artificial structures are viewed and contested (Devine-Wright, 2005; Woods, 2006). Windfarms that work with local community networks are more likely to gain their support and planning permission (Eltham et al., 2008). In other communities, people have made positive associations between wind turbines and an industrial heritage that includes wind pumps or even mine workings (Wheeler, 2017). In these cases, windfarms contribute to a sense of place and wider acceptance, highlighting the need to consider how windfarms reflect or challenge local constructions of rurality.

Based on a study in mid-Wales, Munday et al. (2011) conclude that windfarms offer limited economic and community benefits to local populations. Although the construction and maintenance of windfarms has the potential to create local employment, labour and materials are largely imported from outside rural places, often from overseas. They suggest that only 150 jobs were created through the construction of windfarms on seven sites in rural Wales. Once operational, 'warranty conditions mean that turbine makers tend to use their own staff for on-site maintenance and, with turbine management being largely automated, inspection needs are infrequent' (p. 6). That said, they identify that some commercial companies provided some direct investment to local communities, ranging from £1,000 to £5,000 per megawatt per annum. Some of this was channelled into community groups or buildings, such as sports clubs or village halls, other investments were made to support sustainable energy use, environmental enhancements and habitat management schemes were also initiated, and in a few cases, tourist activities, such as mountain bike trails, were established at the sites of farms. Perhaps unsurprisingly, these mainly benefited communities in the immediate vicinity of windfarms. Munday et al. (2011) conclude that community-owned windfarms had more potential to deliver economic and community benefits. Given these windfarms are in a minority, claims that windfarms, or indeed other forms of renewals, contribute to rural regeneration need to be treated cautiously.

Windfarms illustrate some of the complexity associated with sustainability. Although windfarms contribute to reducing the emission of carbon and the generation of renewable energy, they have little social or economic benefit.

13.4 Resilience

The Covid pandemic, war in Ukraine and severe weather events demonstrate the need for communities to be resilient. Resilience is defined as the capacity for systems to 'bounce back' from shocks or, as Adger (2000: 347) states, 'the ability of groups or communities to cope with external stresses and disturbances as a result of social, political and environmental change'.

The concept draws on both the natural and social sciences (Perrings, 1998). It has been used in maths, physics and ecology to describe respectively how a material can recover its shape following stress, how long a system takes to return to steady state and, in ecology, 'the persistence of natural systems in the face of significant change caused by fires, floods and human interventions' (Robinson, 2019: 56). Within the social sciences, attention has been given to the ways in which societies react and adapt to political, social and cultural shocks (Olsson et al., 2015; Adger, 2000).

Resilience can refer to the restoration of equilibrium between ecological, economic and social systems. It is unclear, however, whether an equilibrium can ever exist and, if it does, what it favours (McManus et al., 2012). It is also implicit that if society 'returns to normal', it remains in danger of similar shocks. Consequently, attention has been given to whether societies can 'bounce forward', using shocks as an opportunity to 'build back better' (Table 13.5) (Scott, 2013). Wilson (2012) suggests that communities can follow different transitional pathways to improve their resilience to future shocks by developing strong economic, social and environmental capital.

TABLE 13.5 Resilience

Equilibrium Resilience	**Evolutionary Resilience**
'Bounce back'	'Bounce forward'
Ability to accommodate disturbance without experiencing change to the system	Ability of a system to respond to shocks by adaptation
Return to 'business as usual'	Opportunity to 'do something different'
Short-term response to shock	Long-term responses
Disaster management and response	Regional, economic and spatial planning
Conservative approach, blames 'nature'	Recognises politics of resilience
Reinforces existing power structures	Critical views, enables reform

Source: Scott (2013)

The idea of resilience is particularly significant to farming as it relies on natural resources that are themselves subject to climatic shocks. Drawing on fieldwork with farmers in New South Wales, Australia, McManus et al. (2012) also identify the significance of social resilience in farming:

> *a sense of social belonging, their engagement with community organisations and their perception of community spirit, have endured despite the decline in the rural population, the ageing of the rural population, and the pressures on farmers due to increased competition, commodity price fluctuations, and drought.*
>
> *(p. 27)*

Agatha Herman also argues that enchantment and emotion – joy, most obviously, but also anger and frustration – bind farmers to places and make them more resilient (2015). Her study of organic farming in Finland shows that resilience also extends to non-human elements of the farm so that 'animals benefit from higher welfare standards, workers and soils benefit from changed chemical usage, farmers benefit from higher market prices while regulators benefit from the influx of certification fees' (Herman et al., 2018: 120). This recognises the 'more-than-human' nature of farming and confirms that resilience should be considered holistically, as something that binds rather than separates 'natural' and 'social' systems.

Yet resilience is a contested idea (Grove, 2018). Walker and Cooper (2011) note that the idea of resilience relies on communities finding their own internal solutions to problems. In this way, resilience has been seen to support neoliberal forms of governance that seek to shift responsibility away from government and towards individuals or communities. Drawing on ethnographic fieldwork with rural communities in Jamaica, Grove (2013) argues that educational programmes and training seek to produce individuals that are aware of risk and can adapt to change in the face of natural hazards. Affective and emotive responses – such as fear and hope – are also incorporated into this training. Thus, the local environment is seen as a source of risk and fear – a river can flood; steep slopes can lead to landslides – rather than reflecting a history of colonial exploitation that has left communities impoverished and vulnerable to shocks. In this sense, resilience is politicised. It is advocated as something that can adapt to natural disasters but not respond to deeper, structural problems in ways that might lead to more sustainable futures. Grove (2018: 269) concludes that

> *risk management becomes individualised and depoliticized: rather than the result of historically specific unequal and unjust political economic relations, vulnerability and insecurity are an inescapable part of the world that individuals are more or less capable of coping with.*

Seen in this way, resilience foists responsibility onto communities and charges them to act in resilient ways to find local solutions to problems that stem from well beyond their territorial bounds (McDonagh, 2014).

13.5 Conclusions

Sustainability is of central importance to rural studies and rural policy-making. Yet, the study of geography is about more than the study of sustainability (Westaway, 2009). As this chapter has discussed, sustainability and resilience entwine economic, social, cultural and natural networks in complex ways. Geographers can recognise Malthusian warnings about climate change but, crucially, can also illuminate pathways towards sustainable living using culturally informed viewpoints, radical politics and, most recently, ideas of hybridity that break down binaries of society and nature. It is therefore crucial to understand diverse rural cultures, nuanced meanings of rurality and how these are played out in different places. The following section examines how the 'cultural turn' in rural geography has sought to understand these issues.

Part IV Cultures

The previous section considered conflict in the countryside. It focused on painting a 'big picture' of change, showing how social, economic and political transition has impacted rural places and people. These changes are by no means one way. Cloke and Goodwin (1992) argued that places are held together by a 'cultural glue' that shapes the direction of change. This is often revealed in local politics or decision making, especially when debates revolve around the different ways in which the countryside is valued (Woods, 2006). Consequently, geographers started to consider the ways in which rural places are 'culturally constructed' and contested. This work sought to identify dominant or hegemonic visions of the countryside and what this revealed about power and class. Attention was also paid to examining different or alternative views of rurality to understand how and why 'other' groups are marginalised or excluded from power.

This became known as the 'cultural turn'. As it gained momentum, ideas of post-structuralism challenged the assumption that rural society was made up of 'structures' or set 'identities' and, instead, that it was continually being defined and reformed. Post-structuralism sought to deconstruct categories such as race, gender and sexuality and examine how meanings associated with these ideas are situated in particular places. The aim was 'to resuscitate the suppressed, to make room for the absent, to see what is invisible, to account for what is unaccountable, to experience what is forbidden' (Gibson-Graham, 1996: 28). As such, the cultural turn has had profound impacts on the way rural geography is studied and the subjects it focuses upon. As this section outlines, it led to growing interests in landscape, hidden others and human-nature relations (Cloke, 1997).

DOI: 10.4324/9780429448966-17

CHAPTER

14

Rural Landscapes

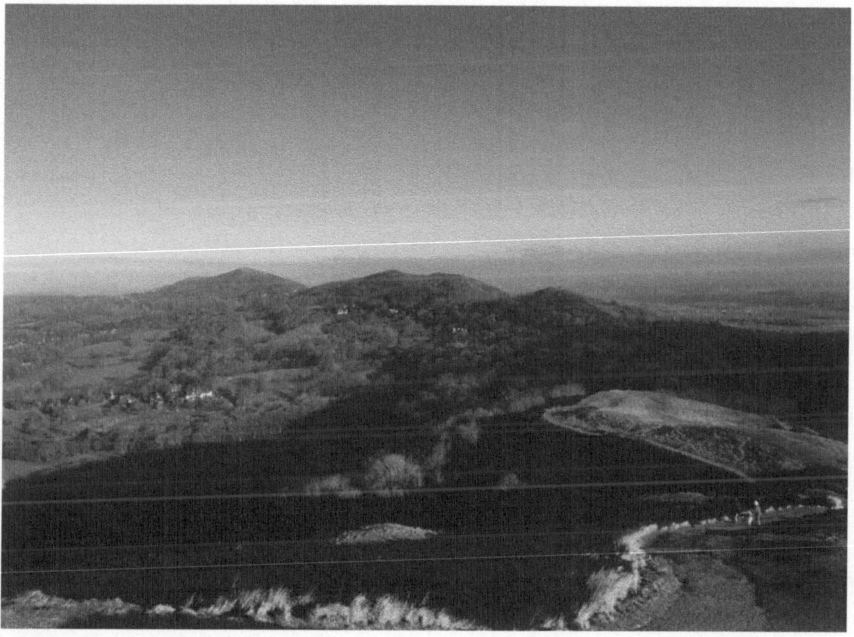

FIGURE 14.1 The Malvern Hills.

The British Camp – also known as the Herefordshire Beacon – that buttresses one end of the beautiful Malvern Hills is one of those happy places where humanity's work has actually managed to improve on Mother Nature's. There is no need to feel guilty here about the enormous disruption our ancestors caused, even to such Arcadian surroundings, when they set about fortifying the second-highest summit of the range sometime in the second century BC.

DOI: 10.4324/9780429448966-18

The Herefordshire Beacon is just a mountain – it stands only 15ft over the traditional 1,000ft (304m) qualification – and it would have been beautiful as a natural, grassy pyramid, but its sinuous earthworks make it sublime. Andy Goldsworthy would die for the sort of budget or – to be accurate – the years of graft by simple people who depended on hand tools – which allowed the sculpture of the entire hill.

The camp itself is part of a British Best View which also takes in the great division of the English Marches: tame but sweetly lovely on the English side; grand and rugged on the Welsh. The first is full of lights at night, the second mysteriously dark . . .

You can see, and count them: Herefordshire, Worcestershire, Gloucestershire, Shropshire, Monmouthshire, Brecknock, Radnor, Montgomery, Warwickshire, Staffordshire, Somerset and Oxfordshire. Actually, we could have a bit of an argument about Oxfordshire, but I think it's there.

On our way to the British Camp, Chris Thomond and I walked through bluebells and milkmaid or lady's smock (don't believe anyone who tells you that British wild flowers have gone), brushed aside orange-tip and green-veined white butterflies, watched a buzzard and a sparrowhawk and avoided an adder.

(*Wainwright, 2010*)

14.1 Introduction

The opening passage describes one of my favourite views, which is from the British Camp on the western end of the Malvern Hills in the English Midlands. The Malverns rise to a modest 1,394 feet (425 metres), yet the relative flatness of the Severn plain extenuates their significance and affords spectacular views of the English and (more distant) Welsh countryside. I grew up and lived for long periods of my life in the English Midlands – Birmingham and then Worcester – and so, on a personal level, the view connects me with a sense of home, familiarity and belonging. I have memories of seeing the hills, exploring the hills with my family and, on one memorable day in my twenties, walking through the mist to come out above the clouds, which I then saw sun-lit below me on the Severn Plain. British Camp, described in the passage, was an Iron Age hillfort built in the second century BC and abandoned by AD 48. The Malverns also inspired Elgar, linking them closely with English national identity. The crests of the hills, as the journalist described, can be ascended by various tracks that not only reveal picturesque views but also bring the climber into closer contact with nature.

As the example of the Malverns reveals, geographers have understood landscape in different ways (Wylie, 2007; Winchester et al., 2013). First, landscape can be 'read' to reveal how various human and physical processes impact places to create unique cultural landscapes. In this account, 'humanity' and 'Mother Nature' are separate but in harmony. Second, the Malverns have been 'imagined' or viewed in particular ways over time – as a defensive site, an Area of Outstanding Natural Beauty (AONB), as quintessentially English or, as in my case, a site of memory and nostalgia. These readings reflect and reveal different power and social relations. Finally, as the account of the ascent implies, people engage with landscapes in physical, emotional and, perhaps, spiritual ways.

14.2 Reading Landscape

Landscapes are often synonymous with rural places. The etymology of the term is derived from Middle Dutch and was used to describe the visual appearance of land and its pictorial representation (Wylie, 2009a). As such, it has become closely associated with rural places, although, of course, its study also extends to urban and wilderness landscapes. The study of landscape was embedded into geography through the seminal work of Carl Sauer (1925) and what became known as the Berkley School of Landscape. Sauer and his colleagues argued that culture imprinted itself on landscape (Solot, 1986; Sauer and Brand, 1931: 2):

> *Culture is the agent, the natural area is the medium, the cultural landscape is the result.*
>
> *(Sauer, 1925: 46)*

> *The works of man [sic] express themselves in the cultural landscape. There may be a succession of these landscapes with a succession of cultures. They are derived in each case from the natural landscape, man expressing his place in nature as a distinct agent of modification. Of especial significance is that climax of culture which we call civilization. The cultural landscape then is subject to change either by development of a culture or by replacement of cultures.*
>
> *(Sauer, 1925: 20).*

Box 14.1 Agricultural Landscapes

Fraser Hart's (1998) study *The Rural Landscape* followed the Sauerian tradition by analysing the significance of landforms, vegetation and 'the structures that people have added' (p. 2) to American landscape. Specifically, he examined systems of land division, fields, industrial structures house types and settlements. Hart also paid particular attention to apparently mundane features of landscape, stating that:

> *every human structure . . . is born to serve some human need and the form of the structure usually reflects its function of the need it was originally designed to satisfy . . . barns, sheds, garages and privies, that are necessary for daily living.*
>
> *(Hart, 1998: 3)*

Echoing Sauer's 'superorganic' concept, Hart referred to culture as 'the beliefs, values, patterns of behaviour, and technical competencies that are learned and transmitted by a group of people. The beliefs and values of different cultural groups have modelled and modified the landscape' (Hart, 1998: 4). He recognised competing and differentiated cultures, the importance of everyday cultural activities (such as skiing for leisure) and, with some humour, that some groups are more influential than others: 'no culture group has had a greater impact on the American landscape

FIGURE 14.2 Agricultural landscape in the USA: distinctive split-rail fencing in 1938.
Source: Wiki Commons

than the middle-class suburbanites who practice lawn worship'! (Hart, 1998: 4).

Throughout the book, Hart examines how physical features, such as river valleys or moraines, and human activities combine to shape the look of the land. His analysis is knowledgeable and informed by local histories and geographies of place. Yet he stops short of reading meanings into or out of landscape:

> *[T]here is no questions that structures can be symbolic, but for whom are they symbolic . . . perhaps it is unwise to start probing too deeply for symbolism that is subconscious at best. When I ask farmers about their buildings, if they give me answers that go beyond mere plausibility and make real sense, I see no need to quiz them further.*
>
> *(11–12)*

Hart's writing is edifying and entertaining, but nevertheless, geographers have since sought to see how different groups of people read the landscape in different ways and how these meanings are reproduced through engagements with the landscape.

Work from the Berkley School followed three key tendencies (Winchester et al., 2013). First was an assumption that culture was 'superorganic', existing above and beyond individuals who were assimilated into it. Second, there was a focus on mapping and cataloguing the significance of social artefacts to different landscapes. Houses, farm buildings, places of worship, field boundaries and so on were charted to examine how different cultures 'diffused' over space. Attention was paid to cataloguing mundane details such as the types of chimneys, roofs or doorways on houses; the construction of fence panels; roofing designs or how crops were planted. Subtle variations in design and practice were taken as evidence of the ways in which different cultures 'evolved' over time or diffused over space. Finally, there was an emphasis on rustic, folk ideologies that were seen to develop a sense of identity and belonging in and between different regions.

Landscape was considered a palimpsest, a document upon which countless layers of human culture had been written. By carefully analysing and unpeeling these layers, geographers could trace how different cultural practices shaped different landscapes over time (Box 14.1). Sauer and his colleagues made an important and longstanding contribution to the study of rural landscapes by emphasising how people adapted to, shaped and changed physical environments (Wylie, 2007).

This way of seeing was often tinged, as Wylie (2007) puts it, with nostalgia and melancholy. These traits were reflected in W.G. Hoskins' (1955) classic text *The Making of the English Landscape*, which traced the landscape evolution of the English countryside and, at the same time, mourned its apparent desecration by modern technologies. Its most famous passage decries:

> *What else has happened in the immemorial landscape of the English countryside? Airfields have flayed it bare whenever there are level, well-drained stretches of land. Poor devastated Lincolnshire and Suffolk! And those long gentle lines of the dip-slope of the Cotswolds, those misty uplands of the sheep-grey polite, how they have lent themselves to the villainous requirements of the new age! Over them drones, day after day, the obscene shape of the atom-bomber, laying a trail like a filthy slug upon Constable's and Gainsborough's sky. England of the Nissen-hut, the 'pre-fab', and the electric fence, of the high barbed wire around some unmentionable devilment; England of the arterial by-pass, treeless and stinking of diesel oil, murderous with lorries; England of the bombing-range wherever there was once silence. . . . Barbaric England of the scientists, the military men, and the politicians; let us turn away and contemplate the past before all is lost to the vandals.*
>
> (*Hoskins, 1955: 298–299*)

14.3 Imagining Landscape

Sauer, Hart and others provided a way of synthesising how diverse human and physical processes combined to shape different, regional landscapes (Cloke, 2004). Their work drew upon rich, ethnographic fieldwork to develop a profound knowledge of different regions (McDowell, 1994), and for Hart (1982), this was considered the 'highest form of the geographer's art'. Winchester et al. (2013) maintain that this approach was ahead of its time as it emphasised the importance of understanding minority groups and exotic influences on rural places.

Yet Sauerian views on landscape were increasingly critiqued in the 1980s. Many felt that an emphasis on rural folk traditions ignored significant social changes in (urban) society and failed to provide a critical analysis of the power structures that shaped society, landscape and inequality. Don Mitchell (2000: 35) reflected:

as American (and British) cities burned in the wake of race riots, the collapse of the manufacturing economy, and fiscal crisis upon fiscal crisis, American cultural geographers were content to fiddle with the geography of fenceposts and log cabins.

In the 1980s 'new cultural geographies' emerged that focused on the different meanings given to a landscape (Cosgrove, 1984; Daniels, 1993). Thus, a field might represent a workplace to a farmer; the outcome of agricultural policy for an official; an opportunity to walk for a rambler; a wildlife haven (or desert) for a conservationist; or a source of food for an animal. Landscape is not, therefore, a product of a 'superculture' but is understood by different people in different ways that reflect dominant power relations in society (Jackson, 1994). The idea that the countryside is idyllic therefore reflects an elite view that is repeated through art, literature, music and other media. Geographers unpacked how different meanings were given to landscapes and, in turn, these might influence behaviour or actions in these places (Table 14.1).

TABLE 14.1 Reading Landscapes

Rule	Explanation	Example
1. Landscape is a clue to culture	Changes in landscape reflect cultural change in society.	The diffusion of cattle breeds reflects changing agricultural practices (Evans and Yarwood, 1995; Tonts et al., 2010)
2. Everyday landscape of common things	Ordinary and mundane features of landscape are often overlooked but can reveal much about everyday cultures; No specific landscape feature is more or less important than others.	A lay-by in a country lane can reveal as much to geographers as a national park.
3. History and landscape	It is important to understand the history of a place to read the landscape.	Indigenous histories of Australia reveal the importance of sacred sites, such as Uluru.
4. Geographical context	Landscape should be understood in relation to other places.	A village landscape should be understood in relation to its agricultural hinterland.
5. Physical landscape	Terrain, climate and geology can be important to people; it is important to understand how these are valued by different societies over time and space (Schama, 1995).	Woods are important in the imagination of many European countries (Dodgshon and Olsson, 2006).
6. Landscape silences	Silences and absences can say as much as what can be seen.	The lack of racial diversity in the portrayal of the English countryside reveals racial exclusion.

Source: Winchester et al. (2013)

Denis Cosgrove (1998, 1985) argued that landscape is a way of ordering the world:

Landscape is thus a way of seeing, a composition and structuring of the world . . . Landscape distances us from the world in critical ways, defining a particular relationship with nature and those who appear in nature, and offers us the illusion of a world in which we may participate subjectively by entering the picture frame along the perspectival.

(Cosgrove, 1984: 55)

Rather taking landscape at face-value, Cosgrove argued that its appearance and representation are the outcome of political conflict. Thus, landscapes are controlled and represented by powerful groups of people for particular ends. As such, they should be read and interpreted in ways that reveal these different power relations:

> *[A] landscape park is more palpable but no more real, no less imaginary, than a landscape painting or a poem. . . . And of course, every study of a landscape further transforms its meaning, depositing yet another layer of cultural representation.*
>
> (Daniels and Cosgrove, 1988: 1)

> *Landscape, and the way it is seen, reflects historical and material processes. Consequently, there are often silences within landscapes that may reflect exclusionary practices.*
>
> (Winchester et al., 2013)

> *In a rural context, the absence of affordable housing reflects the dominance of housing markets and, in relation to these, the interests of property owners.*
>
> (Gallent et al., 2003b)

Reading the landscape in particular ways is important to many rural activities: landscapes are read by farmers as evidence of good agricultural practices (Burton, 2004), walkers learn to interpret terrain from a map (Lorimer and Lund, 2003; Edensor, 2000), and soldiers are trained how to read landscapes for combat (Woodward, 2000):

> *We were taught perspective, to train our eyes and search for a focal point. We had to scan the landscape and pick out the dominant features, just as an artist would peer ever deeper trying to unravel colours. Like the painter who is restricted by the size of his canvas, we were constrained by our arcs of reference. We were also instructed in interpretation: countryside became terrain, rolling hills became gradients that slow down one's progress across country, wild hedgerows became camouflage, mountain streams became obstacles and sources of water.*
>
> (Ballinger, 1992, p. 129, cited in Woodward, 2004)

As Woodward's work shows, military training reflects and develops a masculine gaze that views rural landscape as something to conquer, overcome and control (Woodward, 2000); a stance that is developed through rigorous physical training that pits people against landscape.

This also reflects a historic view that landscape is seen as feminine (Brace, 2003). An example of this is provided in the album 'The Imagined Village' by an eponymous collective of musicians. It starts with a monologue by singer John Copper who quotes his grandfather, Jimmy Copper, talking about the changing landscape of the Saltdean Valley of the South Down hills:

> *It's changed a lot today. When we was up there you used to look down and all you would see is that sheep-cropped turf and the lovely curl of the downs. Those lovely curling contours as it goes down to the sea. It reminded you of a female body. There was a grace and a beauty in these chalk hills 'round 'ere in them days. But now you look down there and all you see is 'ouses, 'ouses, 'ouses.*

The passage, which is clearly nostalgic, shows a love and appreciation of landscape but also reflects a masculine gaze to the extent that the topography of the South Downs is likened to 'a female body' with 'grace and beauty'. It is at once admired and exploited. While farming is seen as adding to the beauty of the landscape – sheep-cropped turf, for example – house-building appears as an intrusion or even assault on the body/landscape.

Edensor (2002: 39) argues that landscape has become a 'selective shorthand' for national identity: 'for the forging of the nation out of adversity, or the shaping of its geography out of nature whether conceived as beneficent, tamed or harnessed'. Thus, he continues, 'Argentina is inevitably linked with images of the pampas: gauchos riding across the grasslands. Morocco is associated with palm trees, oases and shapely dunescapes, and the Netherlands with a flat patchwork of polders and drainage ditches'. These landscapes are also worth defending. Thus, images of the countryside were used in British recruitment posters in both world wars (Bunce, 1994); mountains, which are hard to conquer, symbolise independence in Catalina (Nogué and Vicente, 2004); lakes and forests, defended in the Second World War, are bound into Finland's national identity (Periäinen, 2006); mines are closely linked to Cornish identity (Yarwood and Charlton, 2009); and rivers symbolise real and imagined frontiers between places (McGrath et al., 2020).

At the same time, 'everyday' rural landscapes help to perpetuate social memories (Brace, 1999). For example, the abandoned spoil heaps and jetties in Askam-in-Furness, a former mining village in Cumbria, have become incorporated into local people's experiences of landscape and, as such, play an important role in local identity (Wheeler, 2014). Yet, as the next section shows, there is much debate about what aspects of landscape should, or should not, be valued and how these should be protected.

14.4 National Parks and Other Protected Landscapes

A plethora of designations have been applied to various landscapes to recognise and conserve their distinguishing characteristics (Table 14.2)(Gallent et al., 2008a; Belle et al., 2018). Although these designations often recognise unique or fragile ecosystems, they also reflect the cultural and historical importance of landscapes (Table 14.3) and, as such, are the outcome of social and political decisions.

This is illustrated by national parks, which are perhaps the best known type of landscape designation (Table 14.2). The International Union for the Conservation of Nature (IUCN, 2020) (Table 14.2) defined a 'national park' as 'a large natural or near-natural areas protecting large-scale ecological processes with characteristic species and ecosystems, which also have environmentally and culturally compatible spiritual, scientific, educational, recreational and visitor opportunities', although these criteria are not met by many countries' national parks (Table 14.2). The designation and location of national parks reflects dominant, but contested, ideas about what kinds of landscapes are valued, what kind of activities should be allowed in them and who they are for. Although the 'national' implies that parks are shared and accessible, many citizens are economically, socially or culturally excluded from them. This reflects that national parks uphold certain values.

The idea of a national park was first mooted by the poet William Wordsworth in 1810, when he suggested that the Lake District should be held as a 'sort of national property in which every man has a right and interest who has an eye to perceive and a heart to enjoy' (Whyte 2000: 101). Similarly, in 1832, the American artist George Catlin proposed:

> *great protecting policy of government . . . in a magnificent park. . . . what a beautiful and thrilling specimen for America to preserve and hold up to the view of her refined citizens and the world, in future ages! A nation's Park, containing man and beast, in all the wild [ness] and freshness of their nature's beauty!*
>
> *(cited in Nash 1970: 729)*

Both of these quotations suggest that nature should be appreciated in a particular way by 'refined' people with 'an eye to perceive'.

The idea of a national park became manifest in the USA with the foundation of Yellowstone National Park in 1872 'as a public park or pleasuring-ground for the benefit and enjoyment of the people' (Runte 1997). There are currently 62 national parks in the USA (National Park Service, 2020), but their histories can reflect a white colonization of wilderness. In some cases, indigenous people were displaced from areas designated as a national park to reflect a vision of 'pure' uninhabited wilderness (Spence, 1999; Nash, 1970; Byrne and Wolch, 2009; Cosgrove, 1995). Some European countries also perpetuated an idea of exclusive, white-controlled wilderness when they declared national parks in their overseas territories. In 1915, the Dutch designated the Ujung Kulon reserve in Java; in 1925 the Albert National Park was established by Belgium in the Congo (Nash, 1970). While celebrating and preserving 'wilderness', these parks also reinforced white, colonial power and the exclusion of local peoples.

National parks were established much later in Europe. For example, it was not until 1949 that national parks were established in England and Wales, following sustained campaigning for better countryside access (MacEwen and MacEwen, 1987). Yet, British National Parks are neither national (74% of land in national parks is privately owned, and only 2% is owned by the state [Nadin et al., 2014]) nor parks (as significant numbers of people live and work in them). Consequently, they only meet the ICUN category of 'protected landscape'. They are mainly located in upland areas, reflecting the significance of the picturesque in determining what is seen as worthy of protection, as well as a cultural desire to dominate the landscape by climbing peaks.

Globally, 14.4% of land area and 3.4% of oceans are now covered by a protected area (International Union for Conservation of Nature, 2020). Greater efforts have been made to work with indigenous and local people to manage protected areas and take account of local needs and cultures (Stevens, 2014; Riseth, 2007; Kati and Jari, 2016), yet tensions often remain. In an extreme example, an armed stand-off occurred in Malheur National Wildlife Refuge, Oregon, in 2016 as conservative militia challenged the federal government's authority to manage public lands (Gallaher, 2016; Middeldorp and Le Billon, 2019; Inwood and Bonds, 2017).

More generally, protected landscapes tend to reflect a particular, often elite, vision of the countryside. Discourses of national identity and anti-urban sentiment have the potential to 'exclude ethnic minorities, among other groups, from accessing the countryside, both physically and emotionally' (Askins, 2009: 365). A recent review of protected landscapes in England and Wales noted that

> *most visits are made by the same (better off, less diverse) people repeatedly, and those who miss out are the older, the young – especially adolescents – and those from lower socioeconomic groups and black, Asian and minority ethnic communities*
>
> *(Glover, 2019: 68)*

and that only two out of 1,000 board members were from black, Asian or minority ethnic (BAME) backgrounds. Although our attention is often caught by the spectacular landscapes of protected areas, the 'landscape silences' (Winchester et al., 2013) of who or what is missing are just as telling. If national parks only exist for a narrow group of people, then it should be questioned whether national parks are national at all.

TABLE 14.2 Internationally Protected Landscapes

Ia. Strict Nature Reserve	Strictly protected for biodiversity and also possibly geological/geomorphological features, where human visitation, use and impacts are controlled and limited to ensure the protection of the conservation values	Dipperu National Park, Australia
Ib. Wilderness Area	Usually largely unmodified or slightly modified areas, retaining their natural character and influence, without permanent or significant human habitation, protected and managed to preserve their natural condition	The Imfolozi Wilderness Area, South Africa
II. National Park	Large natural or near-natural areas protecting large-scale ecological processes with characteristic species and ecosystems, which also have environmentally and culturally compatible spiritual, scientific, educational, recreational and visitor opportunities	Guanacaste National Park, Costa Rica
III. Natural Monument or Feature	Areas set aside to protect a specific natural monument, which can be a landform, a sea mount, a marine cavern, a geological feature such as a cave, or a living feature such as an ancient grove	Yozgat Camligi National Park Turkey
IV. Habitat/Species Management Area	Areas to protect particular species or habitats, where management reflects this priority. Many will need regular, active interventions to meet the needs of particular species or habitats, but this is not a requirement of the category.	Pallas Ounastunturi National Park, Finland

V. Protected Landscape/ Seascape	Where the interaction of people and nature over time has produced a distinct character with significant ecological, biological, cultural and scenic value and where safeguarding the integrity of this interaction is vital to protecting and sustaining the area and its associated nature conservation and other values	Snowdonia, Wales
VI. Protected Area with Sustainable Use of Natural Resources	Areas which conserve ecosystems, together with associated cultural values and traditional natural resource management systems	Expedition National Park, Australia

Source: International Union for Conservation of Nature (2020); Dudley (2008).

TABLE 14.3 Rural Examples of UNESCO's World Heritage Sites.

Site Name (Abbreviation)	Country	Year	Main Crop	Property[b] (in Hectares)
The rice terraces of the Philippine Cordilleras (PC)	Philippines	1995	Rice	–[a]
Portovenere, Cinque Terre and Islands (PT)	Italy	1997	Vineyard	4.69
Viñales Valley (VV)	Cuba	1999	Tobacco	–
Jurisdiction of Saint Emilion (SE)	France	1999	Vineyard	7.85
Agricultural landscape of Southern Öland (SO)	Sweden	2000	Mixed	–
Archaeological landscape of first cultivations of coffee (CC)	Cuba	2000	Coffee	81.47
Wachau cultural landscape (W)	Austria	2000	Mixed	18.39
Alto Douro Wine Region (AD)	Portugal	2001	Vineyard	24.60
Fertö/Neusiedlersee cultural landscape (FN)	Austria-Hungary	2001	Mixed	52.00
Tockaj Wine Region historical cultural landscape (T)	Hungary	2002	Vineyard	13.25
Landscape of the Pico Island vineyard culture (PI)	Portugal	2004	Vineyard	190.00
Val d'Orcia (VO)	Italy	2004	Mixed	61.19
Agave landscape and ancient industrial facilities of tequila (AT)	Mexico	2006	Agave	35.02
Lavaux, vineyards terraces (L)	Switzerland	2007	Vineyard	8.98

Source: Gullino and Larcher (2013: 391).

14.5 Embodying Landscape

Although landscape has been strongly associated with visual media, it is not just something that is gazed upon from afar. Instead, we engage with landscape on a daily basis using all of our senses. Edward Relph (1981: 22) commented that landscape is 'anything I see and sense when I am out of doors – landscape is the necessary context and background both of my daily affairs and of the more exotic circumstances of my life'.

Cloke and Perkins (1998) examine the various thrills offered by the 'awesome foursome' in Queenstown, New Zealand. Tourists participate in landscape by falling through it on a bungee cord, flying over it in a helicopter, skimming over it in a jet boat and or being buffeted on a white-water raft. In all instances, a person experiences the landscape physically and emotionally as their body is subject to forces and sensations that are out of the ordinary. Indeed, adrenaline sports – such as surfing, mountain biking, parachuting, tombstoning or extreme ironing – get their names precisely because of their impact on the body, in this case causing it to produce adrenaline in response to perceived danger (Box 14.2).

Box 14.2 Tombstoning

Tombstoning refers to jumping into water from high places and is mainly practised by young people in coastal areas. It is a communal activity that offers young people ways to express themselves through dangerous activities. The term 'tombstoning' was coined by the emergency services to highlight its dangers but has since been appropriated by practitioners to celebrate a sense of risk-taking (Laviolette, 2016). Tombstoning is an embodied practice that incorporates air, sea, waves, wind and coast in the moment of the plunge.

There is a need for tombstoners to understand and engage with the sea in ways that recognise its changeable nature and 'constant state of becoming' (Anderson and Peters, 2014). Knowing where and how to jump can result in spectacular, thrilling leaps that bring the participant into contact, literally, with the sea and coast in unique ways. But failure in this can lead to death and serious injury. Tombstoning is a significant, widely practised but contested activity that is performed at the margins of sea and society. While tombstoning is not illegal, it is discouraged by many authorities, resulting in conflicts between young people and the police.

Equally, more mundane activities also involve bodily engagements with rural landscapes, and in this context, John Wylie (2005) has explored the significance of walking. This extract describes walking up Glastonbury Tor in Somerset, UK:

> *The landscape which sustains the Tor also sustains the climber. Or in other words it makes one a climber. The heaviness one feels in chest and legs is balanced by a growing lightness, a sense of anchorage being slipped, a feeling of occupying an airy volume of depth and of being lightly supported and elevated by the landscape. A double movement occurs, a folding within the landscape like the occlusion of two weather fronts, from which one emerges as a viewer, a seer caught up with the horizon. This reliant subjectivity emerges on the ascent and does not pre-suppose it. Finally, the path twists sharply upward, and the exertion and uncertainty felt upon the steeply shelving steps fix one's eyes to the immediate ground. Head bowed, one reaches the summit of the Tor.*
>
> *(Wylie, 2005: 451)*

The author does not simply view a landscape from the top of the tor. The physical exertion of walking up a hill enrols him into the landscape and, at the same time, affords him particular vistas. This also occurs as he drives to the tor in a car:

> *Into softly hilly country. A series of peaks and troughs, the course of the road through north Somerset, is an affordance. As the car alternately climbs and plunges, so vistas loom and sink. The recumbent fields of the Avon valley, to the east towards Bath, are raised to break into view upon a crest, once, then quickly again, before being sheared off by the planes and angles of the closing foreground on descent. And these vistas slide around the sides of the car, into the narrowing funnel of the visible passed, as the road begins to assume a more definite southerly track.*
>
> *(Wylie, 2005: 446)*

The experience of landscape creates and is created by the motions of the car. Similarly, Wylie (2005) has explored how the process of walking the South West Coast path creates a series of encounters with land and sea. He is at pains – as a picture of his blisters shows! – to emphasise that landscape is encountered and reproduced through bodily experiences. It is not simply something that is gazed at. In this context, Edensor (2000) recognises that walking (re)produces and (re)interprets landscapes through our bodies.

> *[T]he body can never mechanically pass seamlessly through rural space informed by discursive norms and practical techniques. The interruptions of stomach cramps and hunger, headaches, blisters, ankle strains, limbs that 'go to sleep', muscle fatigue, mosquito bites and a host of other bodily sensations may foreground an overwhelming awareness of the body that can dominate consciousness. Moreover, the terrain and climate are apt to impose themselves upon the body, irrespective of discourses about the rural idyll and the Romantic countryside. The body must perform certain tasks, which may be painful or pleasurable in their novelty, or challenging in their awkwardness. Walkers must avoid barbed wire, be wary when passing through fields of bullocks, make sure they do not step in cowpats or mud or in holes, step over logs, leap across streams, negotiate stepping stones and stiles, swat swarms of flies away, avoid brambles, nettles and thistles. These actions dramatically involve bodily actions and reveal physical properties. For instance, climbing over an unstable and swaying fence, the walker may become suddenly aware of the body's mass and weight. Environment and climate thus impose upon walking strategies and sensations. The tactile qualities of many rural paths produce a mindfulness about one's balance as well as a practical and aesthetic awareness of textures underfoot and all around. The walking body treads across rocky ground, springy forest floor, marsh and bog, rough tracks, heathery moorland, long grass, mud, root-lined surfaces, pasture, tarmac and autumnal leafy carpets. Biting insects inhabit long grasses, rain drenches clothes, and frosty air freezes body parts.*
>
> *(101)*

Drawing these ideas together, John Wylie (2007: 164) argues that 'the world is understood to be continually in the making – processual and performative – rather than stabilised or structured via messages in texts and images'. In other words, landscape is more than something that is just represented through the medium of a picture or other text but is made, or brought into being, through our daily lives and activities (Cloke, 2013). In a study of farming in the Orkneys, Jo Lee (2007: 89) concludes:

> *[L]andscape is not just a palimpsest . . . a historical layering in which the present is merely the sum of past episodes, but is also an active, present future-orientated engagement with the environment.*

Non-representational approaches (Thrift, 2008) encourage geographers to consider how landscape's meanings and relevance are continually produced and reproduced through everyday actions, such as farming, driving, climbing, visiting or simply dwelling in places, reproducing landscape (Edensor, 2006; Ingold, 2000).

In doing so, it is crucial to acknowledge that landscapes are not simply understood through the actions of people in them but rather as an assemblage of human and non-human things (Cloke, 2013). Thus, our bodies and landscapes are not 'free floating' or independent of each other but 'emerge through their interactions with each other' (Macpherson, 2010: 3). A climber with the skill to ascend a steep cliff recreates landscape through his or her interaction with the rock – an ability that may escape many of us! The same climb may produce elation or fear so that our emotions shape and are shaped by landscape (Wylie, 2009b). Landscape is experienced through the medium of different technologies. At one extreme, a climber relies on ropes, pulleys, carabiners and the like to do a climb (Barratt, 2012), but something as mundane as a shoe mediates our interaction with ground, mud and stone (Ingold, 2004). Without shoes, walking becomes a more tactile and very different experience (Macfarlane, 2012). Of course, landscape itself is shifting in response to 'natural' or geological processes that, in turn, shape our interactions with it (Massey, 2005) – these, for example, shape the accessibility of the rock to our climber. Equally, our cultural backgrounds encourage us to act in landscapes in particular ways (Couper and Yarwood, 2012). How and where we walk reflects a cultural history of walking for leisure and the role of various agencies, laws and policies in creating or blocking places for us to walk in (Blacksell, 2005; Yarwood 2012).

14.6 Conclusions

Landscape is often synonymous with the countryside, yet it is hugely contested. Geographers have used the idea of landscape to trace the diffusion of different cultures over time and space, as a text from which to read power relations and, more recently, something that is recreated and sustained through everyday activities such as walking or working. The following chapter uses the ideas of performance to examine in more detail how rural landscapes and places are reproduced.

CHAPTER

15

Performing Rurality

15.1 Introduction

Performance is a metaphor that describes 'the ways in which people are predisposed to carry out unquestioned and habitual practices in rural settings' (Edensor, 2006: 484). These performances and the contests that surround them are significant because they help to reproduce rural space, identities and the meanings associated with them. In these ways, the countryside provides a stage for certain performances, while others are kept in the wings or off-stage entirely. Three important but overlapping types of performances are relevant to understanding the countryside: staged, scripted and everyday performances (Woods, 2010, 2011b).

Box 15.1 Performing Rurality at the London 2012 Olympic Opening Ceremony

The London 2012 Olympic Games was opened with a spectacular staged performance that was witnessed by a global audience of one billion people (BBC, 2012). The event, named Isles of Wonder, drew on humour, music and dance to perform a vision of the United Kingdom that celebrated cultural and political achievements, as well as different forms of national identity (Closs Stephens, 2016). Two of the most striking sections were entitled 'Green and Pleasant Land' and 'Pandemonium', which charted how rural Britain was transformed by the industrial revolution. These performances linked rurality and national identity.

The 'Green and Pleasant Land' segment was based on a stage that reproduced a vision of bygone Britain and included a village, fields, a farm, watermill and trees. This provided a backdrop upon which actors played out quintessentially British rural pastimes: villagers danced around

DOI: 10.4324/9780429448966-19

a maypole, men played cricket under a brooding rain cloud, and farmers tended live animals (40 sheep, 12 horses, 3 cows, 2 goats, 10 chickens, 10 ducks, 9 geese and 3 sheepdogs!). Readings and music – including the hymn Jerusalem and Elgar's Nimrod – further enhanced the connection between rurality and national identity.

The vignette presented in 'Green and Pleasant Land' was intended to evoke 'the countryside we all believe existed once . . . the Britain of The Wind in the Willows and Winnie-the-Pooh' ((LOCOG (2012: 20) cited in Baker (2015: 415)). Rather than treating this as an idealised yearning for a rural past that never existed (Mingay, 1989) Danny Boyle, the director of the ceremony, deliberately drew attention to their mythical, yet powerful, status by 'referencing classic novels for children rather than any lived past' (Baker, 2015: 414).

The bucolic scene was deliberately and vividly ripped apart in the 'Pandemonium' section as seven giant chimneys arose dramatically from the (previously) green fields; industrial machinery (including five beam engines, six looms and a crucible) replaced farms; and a larger, grimy, industrial workforce took the place of villagers, all to the accompaniment of mass drumming.

The 2012 Olympic opening ceremony represents one of the most spectacular performances of rurality and national identity. Whilst playing out key features of the rural myth, the event deliberately undermined these through humour, literary allusion and juxtaposition with other aspects of British life. This achieved a 'post-rural' (Murdoch and Pratt, 1993) vision of the UK that simultaneously riffed on established ideas of rurality and identity but, at the same time, acknowledged their mythical status. Its global reach meant that the staged performance was significant, not only for its statements about rural identity but also for its potential to convey these ideas to an audience of millions.

FIGURE 15.1 A Green and Pleasant Land at the London 2012 Olympic opening ceremony.
Source: Mark Pain/Alamy Stock Photo

15.2 Staged Performances

The idea of performance is traditionally associated with an event that is formally staged, such as a play or concert. While performances can take part on literal stages – for example, a folk concert in the village hall – stages can also refer to settings such as 'village greens, farm-life centre, heritage attractions, grouse moors, mountains, long-distance footpaths and footpaths' (Edensor, 2006: 484) that provide a backdrop for daily life. In addition, these stages may be found in other media, including virtual worlds. The computer game *Virtual Farm*, for example, invites players to 'slip into the tractor seat and turn a small farm into a booming business'.

A spectacular vision of rurality was performed at the opening ceremony of the 2012 London Olympics, which presented bucolic visions of the British countryside to a global audience of over one billion (Box 15.1). Smaller but no less meaningful examples are found in rural communities where annual traditions are enacted to give meaning to places and identity to their inhabitants. One example is the 'beating the bounds' ceremony observed in England and Wales, as well as some parishes in the USA and Germany. This ancient, annual tradition involves walking around the boundaries of a parish and using sticks to beat the various stones, monuments and landmarks that mark its boundary. In eras before maps, the practice maintained knowledge of the parish's boundaries and, today, is still performed in some parishes as a way of linking place and identity.

Elsewhere, rituals such as the blessing of wells, May Day dances and other folk ceremonies are kept alive for the same reason. Newer traditions – such as the Burning Man Festival in Nevada, USA, in which an effigy of a wicker man is burnt on the last day – have inscribed and reproduced new meanings on rural places that are now maintained through repeated, annual performances.

FIGURE 15.2 Heritage buildings at Skansen in Stockholm, Sweden.

Rurality, heritage and national identity are performed in the Skansen open-air museum in Stockholm, Sweden. Skasen was opened in 1891 by Artur Hazeliu in an effort to bring together different folk cultures from across Sweden at a time of industrial upheaval and change (Crang, 1999). The site now has over 50 buildings, a farm and a zoo that are populated by staff in period custom who re-enact craft activities, folk customs and farming practices using native animals and livestock. Ideas of social stability, custom and cultural landscapes are woven together through these performances. In doing so, they also reproduce ideas of national identity (Jones et al., 2004) linked to concepts of *hembygd* (home), which refers to a village, region, nation or world, and *bygd* (cultivated land) that relates to dwelling or living. These acted as a counterweight to urbanisation and social change in the 19th century (Buttimer and Mels, 2006). Industrial workers were not represented at Skansen until after the Second World War, although, today, efforts have been made to portray diverse groups of people through stories and events. Skansen is significant because it was the world's first outdoor museum, an idea that has been reproduced in other countries and settings to provide a stage on which to perform 'living histories' of heritage and rurality. Likewise, farm parks, such as Pennywell in Devon, use activities such as tractor rides, milking and animal feeding to play out versions of farming to visitors (from largely urban places).

Music is also a way in which rurality and heritage are performed. Thus, folk (Yarwood and Charlton, 2009), Americana (Halfacree, 2018) and country and western (Gibson and Davidson, 2004) are genres of music that are particularly associated with rural places. Given that this music often draws on influences and styles from around the world, claims of authenticity need to be treated with caution (Connell and Gibson, 2003), but nevertheless, its performance allows stories about rural areas and lives to be brought to wider, sometimes global audiences.

15.3 Scripted

Performance refers to more than a formal stage show. People are encouraged to perform in rural areas by following 'scripts' or accepted forms of behaviour. A clear example of this is tourism, where visitors are persuaded to see things in particular ways, take predictable photos and, in doing so, encourage others to do the same. John Urry refers to this as 'The Tourist Gaze' (Urry, 1990). Where and how tourists should visit a place are determined by various 'signifiers', such as brochures, websites, postcards, tourist signs and blogs that set an agenda about what should be seen and how it ought to be seen. Photographs and electronic media, such as social media posts or Trip Advisor reviews, also contribute to the reproduction of scripted tourist performances (Urry and Larsen, 2011).

These might include particular sites/sights, such as Uluru in Australia or an 'authentic' farmhouse in Brittany. Expected behaviours might include bungee jumping from a bridge in New Zealand, taking a scenic flight across the Grand Canyon or travelling on a mountain railway to enjoy the view from Snowdon/Yr Wyddfa. These signifiers do not only describe places but write the scripts that tourists expect to play out and follow. While experiences and photos seem unique to visitors, they are essentially following the same script as everyone else.

Rural places are curated in order to meet these expectations. Honeypot sites, for example, coerce visitors to popular spots and, once there, manage their visit through parking strategies, signposted walks, pay-to-enter experiences and souvenir shops. As such, tourism is 'a closed self-perpetuating system of illusions' (Urry, 1990: 7) that manages and reproduces tourist expectations. But this is not to say that tourists are dupes: many are happy to go along and enjoy tourist experiences safe in the knowledge that what they are seeing is scripted and far from authentic.

Indeed, many tourist places have been described as hyper-real, offering an illusionary experience of something that has never really existed in the real world. Most obviously, this exists in rural theme parks, such as Peppa Pig World in Hampshire that was praised by British Prime Minister Boris Johnston for its 'safe streets, discipline in schools, heavy emphasis on new mass transit systems'. Until recently, Canadian World in Japan reproduced the places associated with the *Anne of Green Gables* novels set in Prince Edward Island, Canada. It was based in Ashibetsu, a former coal mining town, and was established as part of a regeneration programme to develop tourism. Visitors could see buildings representing Avonlea, watch shows associated with the novels and meet actors dressed as Anne and her friends. This hyper-reality provided an experience of a place without the expense of travel. Sadly, the park closed as visitor numbers fell.

So far, this chapter has focused on events or places that may seem out of the ordinary, such as a holiday or an annual show, yet everyday performances are also significant in reproducing meanings of rurality.

Box 15.2 Scripted Performances: The Tourist Gaze in the Lake District

The Lake District is an upland area in the north-west of England, characterised by dramatic mountainous terrain shaped by glaciation (Massey, 2005). The area is a national park and a World Heritage Site, attracting 15.8 million visitors a year who 'come to enjoy the scenery, peace and quiet and walking but many others visit specific attractions or take part in an outdoor activity' (Lake District National Park, 2022). Knowingly or not, these visitors play out scripts that have been 'written' for them over many years and, in doing so, reproduce cultural expectations and perform the values associated with the location.

John Urry (1995) notes that the Lake District was 'unknown' prior to the 18th century, and its mountains were dismissed by visitors as hideous or frightening. During the 18th century, the Lake District was 'discovered' by wealthy visitors seeking picturesque scenery. Picturesque referred to an emerging form of landscape painting that required a foreground and background, linked by a middle ground. Initially, Alpine scenery provided this – a shack, tree or shepherd might provide a foreground; mountains a dramatic background and a valley, perhaps with a meandering stream, a middle ground (Figure 15.3) – and was often sought by travellers who would paint the scene, perhaps as part of their Grand Tour of Europe. Artists started to visit the Lake District, with its mountains and water, to find and paint the picturesque.

FIGURE 15.3 A picturesque view in the Lake District.

At the same time, William Wordsworth and other poets, such as Coleridge and Southey, began writing about the Lakes, valuing the Romantic and sublime aspects of its landscape (Donaldson et al., 2017). Their writing was inspired by walks (I wandered, lonely as a cloud) that romanticised being in landscape (Cooper and Gregory, 2011). Urry (1995: 197) contends that this led to a 'place myth' being associated with the Lakes seeking a 'semi-private, quasi-spiritual relationship with signifiers of "a nature"'. These themes were also captured by A. W. Wainwright's guides to the Lakes, published in the 1950s and 1960s. These much-loved pictorial guides to the fells espoused solitary walking and the appreciation of landscape (Palmer and Brady, 2007). Ironically, they have inspired millions to retrace Wainwright's steps and climb the 214 summits he identified, which are now referred to as Wainwrights.

Picturesque and Romantic ideas not only led to greater numbers of visitors to the Lakes but, more significantly, a growing desire to conserve its physical and cultural landscapes. In 1810, Wordsworth's *Guide to the Lakes*, for example, proposed that it should be 'a sort of national property' (Whyte, 2000: 101). These ideas influenced the designation of the area as a national park in 1951 and a UNESCO World Heritage Site in 2018. The latter recognised that the cultural and literary associations with the Lake District landscape should be value for their capacity to 'nurture and uplift imagination, creativity and spirit' (UNESCO, 2022).

Visitors to the Lake District follow scripts that perform and reproduce the values associated with the the region. Many, for example, climb mountains to seek solitude, engage with nature and admire picturesque views from the summit. Others visit literary shrines, such as Dove Cottage (Wordsworth's cottage) or Hill Top, home of Beatrix Potter, thereby acknowledging the significance of these writers and their ideas to the Lakes. Despite a long history of physical and social change (Massey, 2005), the Lake District has been regarded as something 'genuinely English', proffering 'a timeless experience of nature, textures of landscape, and cultures; an Englishness defined through a packaged 19th-century Romanticism' (Crang and Tolia-Kelly, 2010: 2325–2326).

Although these performances play to widely accepted, hegemonic scripts, they may be challenged by other performances that do not 'fit' into notions of quiet enjoyment. Wainwright, for example, bemoaned crowds at beauty spots (Palmer and Brady, 2007). Charlie Gere's (2019) provocatively titled book, *I Hate the Lake District*, traces a series of journeys that explore alternative histories of the Lakes that are bound up with, amongst other things, nuclear power, slavery, imperialism, the occult and UFOs. These subtexts are rarely considered in tourist visits.

Macnaghten and Urry (1998: 204) also describe how a walker dressed in 'breeches, brightly coloured socks, orange waterproofs and a rucksack' entered the town of Cleater Moor, an industrial settlement located just outside the national park, and immediately felt out of place. His clothes, while appropriate for a walk in the Lakes, felt like 'fancy dress' compared to the working attire of Cleator Moor's residents: he 'walked out of the Lake District and its particular sense of place and its appropriate leisure practices' (Macnaghten and Urry, 1998: 203-204). The action highlights that the practice of walking is spatially contingent upon prescribed bodily interactions within a particular, imagined landscape.

Troublingly, many of the Lake District's scripts can exclude on the grounds of class, race and gender (Crang and Tolia-Kelly, 2010; Askins, 2006; Tolia-Kelly, 2007). Ideas of fear, terror or awe may be amplified through feelings of racial exclusion (Tolia-Kelly, 2007; Kinsman, 1995; Askins, 2006); isolation and concerns about safety may limit women's participation in physical recreation (Askins, 2009), and people with disabilities may be unable to access high and wild places (Macpherson, 2008).

The Lake District is based on a 'place myth' (Urry, 1995) that is performed by following scripted physical engagements with landscape, usually through hiking. These reproduce ideas of the picturesque and the sublime but, at the same time, hide alternative or different meanings associated with the Lakes. It is, therefore, important to not only consider hegemonic or dominant performances of place but to appreciate how diverse groups of people perform different scripts that challenge how places are understood.

15.4 Everyday Performances

> *There is a lot of tradition, which we respect, and the jackets and trousers are practical as well. They are waterproof and have a lot of pockets, which you need, and the greens and the browns blend in with the landscape. But as for trying to be something we are not, that is rubbish. Well, at least no more than anybody else in the group. We have all got nice houses in the countryside and Range Rovers, and maybe a few horses, but we are not really pretending we are in the landed gentry. And just because a few of us enjoy traditional shooting, we are all still middle class businessmen who have settled down in the countryside with our families.*
>
> *(Heley, 2010: 327)*

In this extract, Jess Heley (2010) reports the words of a wealthy migrant who regards participation in shooting as an important aspect of rural living and, in particular, establishing his identity within the community. His choice of clothes and decision to go shooting reveal a desire to 'perform' local traditions and, in doing so, maintain their importance.

At the same time, the interviewee is conscious of his own identity and acknowledges that, despite his wealth and participation in the shoot, he lacks the status of more established families in the area (the gentry). Instead, the acts of living in 'a nice house', driving a Range Rover and keeping horses are important to his and his peers' identity as wealthy 'middle-class businessmen' and their aspirations for a rural lifestyle.

This passage stresses the significance of understanding how identities are 'performed' or played out in daily life. Crucially, daily routines, such as wearing the certain clothes, driving the right car and conforming to traditions, give meaning to individuals and communities.

Drawing on interviewees to Vilcabamba in Ecuador, Kordel and Pohle (2018) show how migration from the USA is, in itself, an act that plays out (or performs) a desire for a simpler, rural lifestyle. Once the move has been made, migrants from the Global North enact an idealised rural lifestyle based on a healthy way of life. These ideas were made manifest and performed through buying, growing and using organic food. At the same time, many embraced the idea of community spirit and took part in regular events, often based on healthy or mindful lifestyles, to affirm this:

> *The community is really nice. So when we arrived here, we didn't really know anybody and after a month being here, we already had tons of friends . . . There is Yoga every day in town, there is meditation at the meditation centre, there is potlucks pretty much every week.*
>
> *(Kordel and Pohle, 2018: 136)*

The study of 'everyday performance' seeks to understand how daily action and routines reproduce, maintain or challenge ideas of rurality. In doing so, it is also possible to reveal how these performances reflect social structures and power relations. Patriarchal identities, for example, are affirmed through acts associated with or challenge particular gender roles. Thus, caring for children, undertaking administration, running chores and manual work on a daily basis all conform to being 'a farmer's wife' (Ashton (1994) (Chapter 17).

It is important to note that while repeated performances are important in the reproduction of identity, they also help to maintain, defend and contest particular ideals and values bound up with the rural. Unlike a performance at a play or staged events, everyday performances may not follow the same script. Diverse performances serve to reveal how and why the countryside is contested.

15.5 Conclusions

The countryside is not a place that is just viewed from afar by different media but is a place that is lived and experienced through action and performance. Performance is useful for understanding how rural ideas are reproduced through repeated actions. While it is possible to identify dominant or hegemonic performances that reassert commonly held ideas about rurality or a particular rural place, geographers should also be alert to competing or contested performances that reveal how different groups of people see and live in the country. Crucially, as the next chapter discusses, it is important to consider the diversity of these experiences.

CHAPTER

16

Rural Others, Other Rurals

16.1 Introduction

The countryside has been dominated by powerful groups of people whose interests have been reflected in decision making and represented in the way the countryside has been imagined. The dominance of the 'rural idyll' maintains a vision of the countryside that protects property and class interests while excluding others without the means or voice to challenge its hegemony.

Consequently, the interests, lives and concerns of many people have been neglected. This led to a growing discomfort that rural geography has only painted a partial view of the countryside that is, at best, *about* rather than *by* different people living there. As this chapter shows, there have been moves to diversify and decolonise rural geography with the aim of identifying and empowering a much wider range of rural voices.

16.2 Neglected Others

Today's rural landscape has fewer children and fewer places for them. If you see them swarming in part of Wales, you can guess that they are city children from the Outdoor Pursuits Centre. The farmyard itself is no longer a safe place for children . . .

. . . a teacher describes his disillusionment with opportunities for children's play in the South Holderness village where he lived 'huddled in a tight island of buildings, surrounded by a huge hedgeless, treeless plain of fields.'

Other teachers . . . stressed the ingenuity of children in colonizing odd and temporary spaces for play: how the parking area of the sugarbeet factory . . . became a cycle track or roller-skating rink, how straw bales materialized from somewhere for every kind of activity, how ropes and old tyres made swings from overhanging boughs.

Extracts from a 'Child in the Country' by Colin Ward (1988: 100–101)

DOI: 10.4324/9780429448966-20

In 1991, a book review published in the geographical journal *Area* by Andrew Gilg (1991) prompted a debate that was to impact deeply on the study of rural geography. Gilg reviewed Colin Ward's (1988) book *The Child in the Country*, which examined how children played, worked and lived in the countryside, building upon an earlier volume about children's lives in the city (Ward, 1978). Both *Child in the City* and *Child in the Country* drew upon Ward's radical political ideas to argue that children were able to use play and imagination to carve out places for themselves in otherwise adult-dominated environments. Gilg (1991: 90) felt the book was 'a pleasant diversionary read but there is little of substance here, either as a text book or as a provocative piece'.

In response to this indifferent review, Chris Philo (1992) argued that Ward's book deserved much greater attention. Chiming with the wider 'cultural turn' in human geography, Philo (1992: 193) contended that Ward's book 'bristles with thought, imagination and compassion in a manner that surely should excite the attention of anyone concerned with the changing shapes of rural Britain'. He suggested that it drew attention to the diverse and sometimes neglected ways in which different groups of people engage with the countryside. Philo lamented that too often rural geography, and indeed rural society, had concerned itself with:

> *'Metanarratives' which we conventionally relate about the social world inevitably steamroller over the more specific 'stories' that 'other' peoples in 'other' places tell themselves when seeking to make sense of their specific and situated existences.*
>
> *(199)*

While geographers had often been concerned with theorising and researching the big structural changes occurring in the countryside (see Part 2), the different ways that individuals experienced these had perhaps been neglected. Thus, while shifts in productivist farming have been analysed and debated by geographers, we should also consider how these changes affected people bound up in them – perhaps a migrant worker seeking manual labour, a child whose freedom to explore the countryside may be limited by industrial farming, or a family farmer struggling with stress as she finds it harder to compete in global marketplaces.

Philo also suggested that rural studies have been by and about 'Mr Average' or 'white, middle class, middle aged, able-bodied, sound minded, heterosexual men living in major urban centres of the west'[1] (Philo, 1992: 193). By contrast, he continues, 'those who tell the other and often more local "stories" tend not to be such things, and as a result their voices are rarely heard and are even more rarely allowed to qualify (let alone to dismantle) the grand moves of the grand masters' (193). Why not, he argued, listen to the voices of 'other' people who do not fit these categories and have often been marginalised in rural places? So how is the countryside experienced by racial minorities, LGBTQ people or those living with mental or physical illness? Philo demanded that a a person from a minority group groups 'is paid attention, is respected, is treated as equal of the same . . . is granted a voice' (201). He challenged rural geography to address this neglect:

> *[Why] should rural geographers not investigate the social relations of health and illness or of ability and disability . . . and why should rural geographers not reflect upon the social relations of sexuality, and why do they not consider the possibility that their*

> *(as it were) equivalent to the gay and lesbian 'ghettos' and networks described by urban geographers is actually the lack of such phenomena because tightly knit rural communities are such unforgiving sites for the expression of alternative sexualities? And why too should they not think about a multitude of other 'others': gypsies and travellers of all sorts . . . 'New Age hippies' and companion seekers of 'alternative lifestyles', homeless peoples and tramps . . . Where are all of these 'other' human groupings in the texts of rural geography?*
>
> *(Philo, 1992: 202)*

On the one hand, Philo (1992: 199) called for 'a new form of human geographical inquiry open to the circumstances *and* to the voices of "other" peoples in "other" places: a new geography determined to overcome that neglect of "others" which has characterised much geographical endeavour to date'. On the other hand, he stopped short of calling for a paradigm shift, suggesting that such work might complement existing work in rural geography by 'stirring in some additional ingredients into the mix of rural studies' (Philo, 1992: 193), albeit using new theoretical ideas and methodologies.

Philo's paper promoted a response from Jonathon Murdoch and Andy Pratt (1993) that led to an exchange of ideas about the exciting possibilities offered by examining 'rural others' (Murdoch and Pratt, 1994; Philo, 1993). The debate was wide-ranging and constructive but centred on whether Philo's ideas could be taken further. Murdoch and Pratt questioned whether it was sufficient merely to 'listen' to other voices: 'simply "giving voice" to "others" by no means guarantees that we will uncover the relations which lead to marginalisation or neglect. This raises a whole clutch of issues relating to difference, space and power in relation to the "rural"' (Murdoch and Pratt, 1993: 422). Instead, they proposed that greater attention should be given to the ways that particular social relations reflect and reproduce power and powerful ideas. They argue that academics and policy-makers privilege particular definitions of rural and rural society in ways that exclude 'other' viewpoints, peoples or perspectives. They concluded that 'rather than trying to "pin down" a definition of rurality or the rural, we should explore the ways in which rurality is constructed and deployed in a variety of contexts' (423). By urging scholars to consider the effect of different constructions of rurality, Murdoch and Pratt also challenged them to consider how their own knowledge of the countryside reflected dominant ideas.

16.3 Rural Others: Some Examples

Philo's intervention was embraced enthusiastically by many rural geographers, prompting a host of studies on rural 'others'. Examples include work on race (Chakraborti and Garland, 2004; Kinsman, 1995), gender (Little, 2002a), age (Harper, 2005), youth (Panelli et al., 2002a; Leyshon, 2002), poverty (Cloke et al., 1995a), religion (Tonts, 2001), sexuality (Smith and Holt, 2005; Bell and Valentine, 1995; Little and Leyshon, 2003) and physical and mental illness (Parr et al., 2004; Philo et al., 2003). Three edited books (Cloke, 2003; Cloke and Little, 1997; Milbourne, 1997) captured the diversity and excitement of work at the time and re-enforced arguments for taking a 'cultural turn' in rural studies.

A significant body of work emerged on alternative practices and countercultural movements, sometimes revisiting well-known ideas, such as counterurbanisation, to reveal hitherto unnoticed trends (Halfacree, 2006, 2009). Other work provided new perspectives on groups that had been regarded as powerful. Thus, issues of farming stress were brought to the fore, dispelling the idea that farmers were tough and, instead, revealing their vulnerabilities in light of social and economic change (Jacob et al., 1997; Meyer and Lobao, 2003; Price and Evans, 2009).

The aim of this work was to offer a view from 'within' the countryside, rather than from the 'outside looking in'. This led to new research methods aimed at capturing different voices, especially those of lay people. Different media, including film (Fish, 2007), photography (Kinsman, 1995) and music (Yarwood and Charlton, 2009), were utilised in this quest. Even quantitative methods, such as questionnaires, were revalued for their ability to reveal minority viewpoints (Cloke et al., 1997). Given this diversity, it is not possible to examine every avenue of the cultural turn, and, instead, the following sections are used to exemplify the value of this approach.

Rurality and Race

The significance of race and ethnicity has been overlooked in rural places (Panelli et al., 2009; Chakraborti and Garland, 2013). In England, 98% of those living in rural places are from a white ethnic group (Department for Environment, Food and Rural Affairs, 2022) (Table 16.1), and racial minorities are less likely to visit the country than white people (Glover, 2019). Given the absence of black, Asian and minority ethnic (BAME) people in rural areas of the West, it was wrongly assumed that race and racism were not significant issues in the country. Agyeman and Sooner (1997: 197) commented that 'ethnicity is rarely an issue associated with the countryside. Its whiteness is blinding to the presence in any other form than the non-white'. Although racial conflict or segregation may be more overt in urban areas, the lack of racial minorities living or using the countryside is itself indicative of racial exclusion (Cloke, 2006b). This is so much so that in 2004 the chair of the UK's Commission for Racial Equality, Trevor Phillips, described it as a form of 'passive apartheid' (Neal and Walters, 2006; Chakraborti, 2010).

TABLE 16.1 Ethnicity in Rural England, 2019

Ethnic Group	Population	Percentage
White	9,587,100	97.9
Asian/Asian British	97,000	0.98
Mixed/multiple ethnic groups	93,000	0.94
Chinese, Arab and other ethnic groups	55,300	0.55
Black/African Caribbean/Black British	41,200	0.42

Source: Department for Environment, Food and Rural Affairs (2022)

In part, this division reflects economic inequalities between races. People from BAME backgrounds are under-represented in the middle and service professions that have the wealth and opportunity to move to rural places (Agyeman and Sooner,

1997) (Chapter 8). Economic disadvantage, which is more likely to be experienced by BAME people, also reduces opportunities to visit rural places for leisure and recreation (Glover, 2019). Racial inequalities in the country are therefore an outcome of racial and structural disadvantage in wider society.

This exclusion is exacerbated further by a cultural narrative that presents the countryside as 'white space' (Agyeman and Sooner, 1997; Neal, 2009). This was vividly illustrated by Ingrid Pollard's photographs titled *Pastoral Interludes* that were first exhibited in 1984. These photographs showed black people, including the photographer herself, visiting the Lake District and other rural places (Kinsman, 1995). The pictures are striking because they are both ordinary and extraordinary. They are ordinary because they show typical rural activities – such as rambling, fishing or enjoying views – but, at the same time, are extraordinary because they show racial minorities undertaking them. They are striking because, at the time, few black people appeared in images of tourism in the Lakes. As Kinsman (1995: 306) comments, Pollard's 'position in this relationship is ambiguous, as the landscapes she photographs have been an environment where she has experienced being the object of intense observation, and yet she becomes an observer herself'.

The accompanying text juxtaposes Wordsworth's poem 'Daffodils', often associated with an idealised countryside and national identity, with her own fears of visiting the countryside:

> *It's as if the black experience is only lived within an urban environment. I thought I liked the Lake District, where I wandered lonely as a black face in a sea of white. A visit to the countryside is always accompanied by a feeling of unease, dread . . . feeling I don't belong.*
>
> *(Quoted in Kinsman, 1995: 301)*

Both the photographs and text confront racialised images of British rurality by showing visiting the countryside from a black perspective. The pictures starkly portray that the countryside is often portrayed as 'white space', a dominance that is reinforced through hegemonic images and, sometimes, overt forms of racism that associate ethnic minorities with urban space and regard their presence in rural areas with suspicion or hostility (Holloway, 2007; Cloke, 2006b; Neal and Agyeman, 2006). They reveal the importance of listening to the experiences of minority groups in rural places and the power of using visual images to bring these issues to a wider audience

Although ethnic and racial minorities are often excluded from representations of rural space, it is important to recognise that these people have had a long association with rural space (McAreavey, 2016; Bressey, 2009). Migrant workers, for example, are a significant presence in rural places, although they are often out of sight if they are accommodated away from resident populations (MacKrell and Pemberton, 2018). Similarly, practitioners of Islam may appear hidden in rural areas, especially if worship occurs in private or public buildings that have other purposes (Dafydd Jones, 2010).

In many countries, the histories and geographies of rural space are closely associated with indigenous people who lived in them prior to colonisation. Yet, these groups have been socially, politically, culturally and economically marginalised by colonialisation and its legacies. In Australia, for example, native people were decimated by disease and driven to marginal places by settlers and farmers. During the 20th Century, they were exploited for cheap labour and divided by egregious policies that forcibly removed mixed race children from their parents. There were also restrictions on movement, often confining indigenous people to places outside 'white settlements' and subjecting them to racial policing (Carter and Hollinsworth, 2009). Indigenous Australians were not granted full political rights until 1967. They continue to suffer de facto exclusion compared to white Australians, as evidenced by higher mortality rates, higher unemployment, lower educational achievement and higher rates of drug and alcohol addiction (Tonts and Larsen, 2002). The Northern Territory National Emergency Response was a series of measures imposed by the federal government (2007 – 2012) to tackle these issues. It resulted in a series of controversial measures, including the suspension of some racial equality laws, that questioned indigenous communities right to govern themselves. These examples demonstrate that although populations of indigenous people are significant in many rural places, they are marginalised in ways that disempower, neglect and hide their contribution to rural society. Similar issues occur in countries that have been colonised by Europeans (Ip, 2003) and, as Ingrid Pollard's work highlights, many people of colour have been displaced and marginalised by the slave trade and its legacies.

Academic study has also often overlooked 'major historical and contemporary forms and trends .... outside the metropolitan centers of the northern hemisphere' (Carrington et al., 2015: 3). Often narratives have been written by the colonizer, rather than the colonized. In part, a growing awareness of the global countryside is challenging these conventions (Woods, 2007), especially when work is aimed at exposing unequal, transrural relations between places (Askins, 2009). More significantly, there have been moves to decolonise the study of geography and destabilize categories of race (and other social structures) in order to empower those who have not had the power to be heard (Connell 2014). This has challenged the hegemony of academia in the Global North and led to calls for research to be by, rather than of or about, those from the Global South (Panelli et al 2009).

Indeed, as work has progressed, race, ethnicity and indigeneity have been understood in more nuanced ways (Panelli et al., 2009) (Jordan et al., 2009). Kye Askins' (2009) work on the experiences of different ethnic groups visiting English national parks reveals how visitors related the sights, smells and sounds of the countryside to places they were familiar with overseas:

> *I mean OK so you don't haven't got lions and hippopotamus and giraffes like in parts of Africa . . . umm but it's the same thing . . . animals in the wild and open views and all of that it's all countryside.*
>
> *(Individual interview in Sheffield: man, 35–44, Ghanaian)*

> *[I]t's the same countryside here and in India. OK it looks a bit different our hills are. They have proper mountains here we have hills but it feels the same.*
>
> *(Individual interview in Middlesbrough: woman 25–34, British Asian).*
>
> *(Interviewees quoted in Askins, 2009: 371)*

FIGURE 16.1 Produce from 'The Black Farmer' celebrating black history.
Source: The Black Farmer.

Rather than seeking to fit into an English rural idyll, their experience of the countryside is a hybrid mixing of places and times. Askins use the term 'transrural' to describe how 'rurality implicates other spaces and places, not only with regard to its binary "the urban" but also networks of spaces and places across different scales' (p. 373).

One example of this is provided by Wilfred Emmanuel-Jones who is also known as 'The Black Farmer'. Born in Jamaica, Emmanuel-Jones grew up in inner-city Birmingham (UK) before establishing a successful business career in television and marketing that allowed him to buy a farm in Devon (Emmanuel-Jones, 2023). The farm was located in an area with very little ethnic diversity, which prompted him to draw on his racial identity to brand his produce:

> *People thought I was a drug dealer and I'd bought my farm with money from drugs. The stereotypes are all around. What I did with my brand is make a virtue out of my colour.*
>
> *(National Union of Farmers, 2022)*

Emmanuel-Jones established 'The Black Farmer' brand to market a range of meat products (Dinkovski, 2022) (Figure 16.1). The name 'Black Farmer' was chosen because it had a sense of 'jeopardy' or, as Emmanuel-Jones explains, 'an extra edge, because people might not be sure whether it was politically correct or not' (quoted in Caines, 2022). This careful branding celebrated his racial identity and invited the public to do the same by purchasing his products. In this way, the 'Black Farmer Brand' contributes a re-imagination of race and rurality that celebrates diversity.

Social movements have also been significant in highlighting racism in rural places. The Black Environment Network (BEN) was co-founded by geographer and activist Julian Agyeman (Agyeman and Sooner, 1997) and campaigns on issues of environmental justice and race, including wider participation for ethnic minorities in rural areas. In 2020, the Black Lives Matter (#BLM) movement mobilised to fight against racial injustice and exclusion at a global scale. Although cities were the main foci for these actions, activism also occurred in rural places (Box 16.2).

Box 16.1 Black Lives Matter: A Rural Perspective

The Black Lives Matter (BLM) movement was established in 2013 in the USA by four black women following the fatal shooting of Trayvon Martin in Sanford, Florida. Martin was a young black man who was killed by George Zimmerman, a white, unofficial neighbourhood-watch coordinator, who confronted, pursued and then shot him following an altercation outside a grocery store. Zimmerman was charged but acquitted of murder and manslaughter after the jury accepted that he acted in self-defence (Yarwood, 2020b).

Black Lives Matter started as a social media campaign to protest about the outcome of the trial and, more widely, draw attention to racism in the judicial system (Derickson, 2017). With the deaths of more black people at the hands of police officers, the campaign gained momentum. It crystallised at protests across many cities in the US and the emergence of a global network of activists in many countries (Black Lives Matter, 2022).

On 25 May 2020, George Floyd was killed by a police officer who knelt on his neck for nine minutes despite pleas from Floyd that he could not breathe. Widespread protests occurred across the world, challenging many racist histories and assumptions about public space. Public statues were defaced or destroyed and the names of public places were challenged for their associations with slavery or colonialism.

Although most protests occurred in urban places, the BBC (2020) reported how attempts to organise a protest in the Forest of Dean, Gloucestershire, met with resistance. In response to years of racism and exclusion, Khady Gueye, a young black woman, organised a BLM pro-

FIGURE 16.2 Khady Gueye with the art installation Soil Unsoiled in the Forest of Dean.
Source: PA Images/Alamy Stock Photo

test in her hometown of Lyndey with her friend Eleni Eldridge-Tull. This occurred during a period when the Covid-19 pandemic placed restrictions on public gatherings, but permission was given by the local council and police for the event. A petition against this was organised by a local resident that sparked debate on social media. The council initially rescinded their decision on safety grounds but used the term 'all lives matter' on various occasions, a term that has been associated with racist groups. The council later apologised for using the term and reinstated the rally, which went ahead peacefully in June 2020. The organisers, though, received threats, and one councillor resigned and left the area following the abuse she received for attending the event (BBC 2020). A few months later, Guaye collaborated with local artists to produce a poem and sculpture that examined the experiences of racial minorities in the area, but this, too, was met with hostility in some quarters (Figure 16.2) (BBC 2020). Gueye and Eldrdige-Tull established the Local Equality Commission with the aim of advocating racial and economic justice in rural areas by working with rural communities and educating young people.

Ability/Disability

FIGURE 16.3 Access to the countryside for sight-impaired people.

The countryside is often viewed from the perspective of those who are able-bodied or sound-minded (Philo, 1992). To redress this, geographers have examined how people living with physical or mental illness experience and use rural places. Hannah Macpherson (2008b, 2009) revealed how a walking group for visually impaired and blind people challenged associations between landscape, leisure and visual appreciation. Recounting one walk, she notes:

> *[W]e sit and drink water as landmarks are pointed out and rare grasses passed around. My companion feels the grass and jokes about never having been interested in these things even when she could see them. Another visually impaired participant, who has been a rambler in the Peak District for over forty years, takes out a monocular to view the distant hills with his fading sight. We continue up the hill and I pick out a way through the increasingly rugged terrain, our route structured by my sight and my attempts to avoid gulleys and find paths wide enough for two. As we ascend arm in arm the peat resonates beneath our feet.*
>
> *(Macpherson, 2009: 1043)*

Touch – feeling the grass, avoiding gullies, the feeling of walking on peat – is essential to the way that landscapes are co-produced and experienced by both walker and guide. Tactile rather than visual relationships with landscape assume greater significance as walkers negotiate a path. At the same time, Macpherson also notes that a 'historical relationship with one's own embodied past' (1049) is also important. Blind people often drew on their memories of seeing landscapes and combined them with descriptions from their guides to build a picture of a place:

> *Ellen: 'Well you must remember of course that I have got years of memory, and I may see an outline of hills and then use my monocular to give a bit more detail and depth to the hill. I like big spaces obviously, looking out across mountains and big valleys . . . and people will say like oh there are three ridges there, one behind the other, and I can envisage that. I don't see it, but I can envisage it because I have got a great deal of visual memory'.*
>
> *(Macpherson, 2009: 1049)*

Humour also played a strong role in these engagements (Macpherson, 2008b): laughter helped to forge bonds with guides, liberate blind people from a sense of pity and equip them to deal with setbacks or prejudice. It is a reminder that people engage with landscape both emotionally and physically.

Parr et al. (2004) have revealed a rural setting can affect the way that people living with mental health are treated. Drawing on interviews with people living with mental illness in remote Scottish locations, they revealed a range of responses to their illness. Some were treated as 'eccentric locals', others as suspicious newcomers. In some cases, people were supported by friends and neighbours through visits and helping out with errands:

> *The neighbours I had in those days . . . [had a] caring and wonderful way of realising my situation . . . [they] were so understanding, so really, really understanding. Wherever they could, they would help me along. The post office was miles away and we all need a post office at times, or I had to see a doctor in Inverness, or food: all sorts of things, they were absolutely marvellous. I couldn't have survived without them.*
>
> *(Kyla, Inv., 1 June 2002; quoted in Parr et al., 2004: 406)*

Others reported an awkwardness by other people at their situation:

> *It's like, if somebody's broken a leg, you immediately go to their house and help them as much as you could and make them meals, make them comfortable, make them cups of tea,*

and this sort of thing. You would do that as a good neighbour, just automatically. But with mental health problems . . . it's difficult, it's strange.

(Darren, NWS, 18 July 2001; quoted in Parr et al., 2004: 406)

At other times, people found themselves the butt of jokes, avoided in the street or simply met with silence. In some cases, though, this escalated into a hostility that was exacerbated by the visibility of rural living. This made it difficult to live privately or escape the gaze of other people:

But we got about halfway down the High Street and there was these two women standing blethering and pointing at me, you know. And that was it, I just had to go back home. I couldn't continue to go down the street.

(Charmaine, ER, 22 November 2001; quoted in Parr et al., 2004: 407)

Philo et al. (2003: 263) go on to stress that greater attention is needed on the daily 'life-worlds' of people living with disability, or

the everyday lives of people with both incipient and diagnosed mental health problems as they struggle to cope with the everyday hassle of living, maybe working and possibly 'playing' (spending leisure time) in rural areas.

This idea was developed in a special issue of the *Journal of Rural Studies* (Pini et al., 2017), in which a series of articles explore how rurality was experienced by people with various forms of disability, including people recovering from strokes (Meijering et al., 2017) and with learning disabilities (von Benzon, 2017; Bryant and Garnham, 2017). Crucially, all of these accounts aim to make disability *visible* in the country and, in doing so, overcome some of the ablest views of rural life (Cella, 2017).

This work also draws away from a medical view of disability that regards it as something that hinders a person and, instead, takes a social approach that realises disability reflects how society, and in this case rural society, regards (dis)ability. Seen in this way, inclusion and exclusion reflect societal structures and practices rather than disability per se (Butler and Parr, 1999).

16.4 Intersectionality

A focus on 'other groups' did much to diversify rural studies and draw attention to a wider range of people, voices and experiences. At the same time, Cloke and Little (1997: 11) questioned whether a rather banal 'off-the-hip list' of marginalised groups had emerged, perhaps to counter Philo's Mr Average. They went on to caution that defining otherness on the grounds of 'gender, sexuality, race, age, disability, alternativeness and so on' risked oversimplifying complex issues and relationships. Such approaches might also miss 'other others' who remain unseen and unheard, sometimes in plain sight. Farmers, for example, are often assumed to be one of the most powerful groups in the country, yet work has revealed their economic and emotional vulnerability (Kelly and Yarwood, 2018).

In an effort to overcome these issues, geographers have considered the significance intersectionality (Hopkins, 2018). Intersectionality recognises that social characteristics, such as class, race, sexuality or gender, are not independent of each other but, rather, are 'mutually transformative and intersecting, each altering the experience of the other' (Ruddick, 1996: 138). The idea was developed by black feminists in response to work that overemphasised the experiences of white middle-class women. By contrast, they argued that black women were not only oppressed by their gender but also by their race, class and other circumstances (Crenshaw, 1990). Thus, work that examines the link between poverty and gender can mask higher rates of poverty experienced by black women; at the same time, demands for racial equality may privilege the interests of black men and therefore continue to neglect black women (Norris et al., 2010). It is, therefore, important to consider social categories in relation to each other and to appreciate their historical and spatial contingency (Walker et al., 2019). Specifically, Hopkins (2018) argues attention should be given to the following:

1. social inequality between different groups of people;
2. the power relations that shape these inequalities;
3. relationality – how inequalities are manifest in relation to others;
4. the social context – including how places reflect and affect inequalities;
5. complexity – how these issues are brought together in different ways in different times and spaces;
6. social justice – how work can lead to change and more equal social relations.

Table 16.2 exemplifies some of the ways these issues can be operationalised in rural studies (Walker et al., 2019). At the same time, it is important that geographical research remains 'open, exploratory and provides respondents with an opportunity to share parts of their lives that the researcher may not necessarily have considered significant' (Hopkins, 2018: 588). Thus, a project examining exclusion in rural areas might focus on gender and class, but for some people, the intersection with perceived status as 'insider' or 'outsider' may be more significant in their daily lives.

TABLE 16.2 Conducting Intersectional Research in Rural Geography (After Walker et al., 2019)

Intersectional Attribute	Description	Guiding Questions
Intersecting social categories	Social identity categories (gender, class, race, ethnicity, age, etc.) do not operate distinctly from one another but intersect and co-constitute one another to produce context-specific experiences.	• Which social categories are present in our research context? • Which social categories most significantly shape people's experiences in this context? • How do these social categories interact to mitigate and/or exacerbate their respective effects? • Which categories might we have missed? • What do the experiences of research participants tell us about the enduring relationships and neglected intersections producing inequality?

Multi-level analysis	Differential experiences of and responses to policies and practices shape and are shaped by social structures and representation.	• How do experiences across intersecting identities interact with social structures, norms, ideologies and discourses to influence differential access to resources and decision-making for immediate response and long-term adaptation planning?
Power relationships	Multiple systems of power (e.g., sexism, racism) operate together across levels of society to influence the exclusion or inclusion of certain knowledge, experiences, and access to material resources. These inequalities can be reinforced or challenged through institutional, political, and sociocultural practices, discourses and norms.	• How do generally vulnerable societal groups experience climate hazards? Generally privileged groups? What is the relationship between these? • What types of environmental and social knowledge are recognised and privileged in immediate responses and adaptation planning? • Which experiences and knowledge are not reflected in immediate responses and adaptation planning? What are the underlying assumptions that result in these exclusions? • How do immediate responses and adaptation planning either reinforce or challenge social inequalities? • In the context of the study, how has power stabilised to form general vulnerabilities or privileges at specific intersections of identity?
Reflexivity	Critical reflection on researchers' own social positioning and assumptions within broader systems of power and historical, institutional and cultural contexts facilitates research that avoids reinforcing existing social inequalities within the context of the study.	• What are the historical, institutional and cultural contexts in which the research communities are embedded? • What are the researchers' own assumptions about which experiences and knowledge are valuable for immediate response to climate hazards and long-term adaptation planning? • How do the researchers' social locations shape the information shared by participants? • How do differences in power and privilege shape research relationships? How can/did the researcher facilitate power-sharing?

16.5 Conclusions

This chapter has used case studies of race and disability to demonstrate the value of Philo's (1992) call to study neglected rural others. His ideas have been applied to understand many other groups of people and provide new perspectives on rural life (Cloke and Little, 1997). Interest in rural others had a profound effect on rural geography and continues to influence how it is studied. More widely, the importance of diversity has now become incorporated into mainstream thinking within and outside academia – many organisations undertake diversity training, and over recent years, the media has sought to diversify the range of people portrayed in key roles.

Like any paradigm shift, though, there have been questions and debates about its significance (Miller, 1996a; Cloke, 1996). In 1997, Paul Cloke (1997) wrote an editorial in the *Journal of Rural Studies* celebrating the 'excitements' generated by the cultural turn, but, by 2006, he struck a bleaker tone when evaluating its legacy (Cloke, 2006a). He suggested that, in many cases, rural geographers had either ignored the cultural turn or only adopted it in a half-hearted way. Some, he argued, equated it with other areas of geography, such as urban studies, and had critiqued it as self-indulgent or divorced from rural policy or planning. For example, Keith Hoggart (1998: 384) lamented 'The gigantic leap from exotic, high-powered theorisations . . . and the too regularly little-more-than-descriptive accounts that follow', and in another exchange, Simon Miller (1996a, 1996b) questioned the analytical value of the cultural turn.

Further, Cloke and Little (1997) cautioned that studying other groups may represent a form of tourism or voyeurism with a tendency to 'flit in and flit out' (p. 11) of other people's lives in the pursuit of academic fashion. Despite efforts to listen to and empower minority groups, the study of or with rural others is largely undertaken by privileged academics, like myself, who fit the Mr Average bill. In response to these concerns, geographers have paid more attention to their own positionality and backgrounds when studying other groups of people. Work by feminist geographers, for example, has encouraged us to consider the power relations of research. How do research subjects view researchers? Do they feel empowered or intimidated by researchers asking about their lives? How can researchers make studies with, rather than of, diverse groups? In other words, a study of 'others' also requires a consideration of ourselves and the ways our backgrounds, ideas, beliefs, politics, age, gender and so on impact how we relate to people. Positionality determines how different voices are heard or reported with significance for the way research is reported (Chiswell and Wheeler, 2016).

The call to study rural others was a significant moment in rural geography. It widened the scope of rural geography and opened the discipline to new methodologies and new, critical ways of thinking. Studies placed neglected groups on the centre stage of rural geography, highlighting concerns in their lives and, sometimes, empowering them to overcome some of the disadvantages they faced.

Note

1 Assuming Plymouth is a major Western urban centre, I fulfil all of these categories.

CHAPTER

17

Gender, Sexuality and the Body

17.1 Introduction

Gender plays a significant role in rural society (Little, 2002a). Women and men are expected to behave in ways that conform to particular expectations about rural and agricultural living. Geographers have been slow to acknowledge these differences, in part because the subdiscipline has been dominated by men (Leyshon, 2002). Yet, women are under-represented in rural politics; have fewer social and economic opportunities than men; and are often hidden by dominant, masculine ideas of rurality (Little and Panelli, 2003). Since the 1980s, more work has sought to expose and challenge some of the systematic disadvantages faced by women in the country (Little, 1987; Little and Morris, 2005; Pini et al., 2020). This change reflected a wider paradigm shift in geography that drew upon feminist perspectives to highlight the significance of gender to society and space (McDowell, 1983, 1992; Massey, 1994; Women and Geography Study Group, 1984).

Given rural geography's traditional emphasis on agriculture, farming provided the initial focus for feminist research (Whatmore, 1988; Brandth, 2002), but over time, other studies examined women's experiences in different sectors of the rural economy and, more widely, rural society. Thus, feminist geographers drew attention to the ways in which women's lives could be limited by poor transport and limited services (Fernando, 1998). Others paid greater attention to private space, revealing hidden issues of domestic violence and assault (Panelli et al., 2004). These studies drew on different ideas of rurality, noting that women's lives were not only limited by distance and isolation, but the ways in which they were expected, and indeed coerced, to conform to cultural expectations of the countryside (Little, 2002a). While women's issues continue to be an important aspect of gender studies, other work has examined masculinity and sexuality (Campbell et al., 2006), widening the scope of study to recognise diverse gender identities.

17.2 Farmer's Wife?

In 1999, *Country Life* magazine ran a feature called 'The Farmer Wants a Wife', in which single male farmers advertised for a wife (Little, 2007). The idea went on to

DOI: 10.4324/9780429448966-21

become a popular TV reality show that aired in over 30 countries (Stenbacka, 2011; van Keulen and Krijnen, 2014; Peeren and Souch, 2019). In the original magazine series, farmers sought spouses:

> *who had a clear comprehension of what it meant to run a farm. The requirement was not necessarily for someone who had the scientific or technical knowledge to contribute to the farm business, rather it was for someone who appreciated the difficulties of running a farm and could understand that farmers' had to prioritise the farm over a relationship.*
>
> *(Little, 2007: 411)*

Notwithstanding the scripted nature of reality TV, 'Farmer Wants a wife' reveals that farming is more than an economic activity and is associated with particular identities, practices and cultures (Morris and Evans, 2004). Roles for men, women and children are defined and reproduced through farming activities and maintained through gendered practices.

The family farm is both a place of work and a place of home that promotes and relies upon an expected set of gendered, heteronormative work and family relations (Whatmore, 1991a; Price and Evans, 2006; Riley, 2018; Argent, 1999). It positions the man as head of the farm enterprise with a 'farmer's wife' as his subordinate (Brandth, 2002). She is expected to undertake domestic tasks and also play key roles in running the farm whilst, at the same time, excluding herself from decision-making and ownership. Bryant (1999: 247) quotes an Australian female farmer:

> *We pick the banksias once a week. It is my job to pack them and oversee the picking. My husband looks after the yabbies, pine trees for post production, the fat lambs, wheat, oats and other crops. The banksias are really what I do and manage but Len did set it up and tends to do the marketing and make decisions about technical things like irrigation. He looks after the farm as a whole, he does the management and I do the books. The first and last decisions stay with him. He keeps all the different enterprises going and makes sure the whole thing works.*

Romanticised images of the 'farmer's wife' suggest that a woman's role is confined to cooking, housework and lighter farming tasks such as collecting hen's eggs (see Figure 17.1), but work by feminist geographers suggests that this is far from the case. Pioneering work by Sarah Whatmore (1991b) showed that women regularly undertake hard manual work, as well as undertaking bookkeeping and farm administration. By way of example, Sian Ashton (1994: 124–125) recounts the day in the life of a farmer's wife in North Wales in a 'quiet' period:

> *Up at six, breakfast for the family (six in all), do dishes and put laundry in machine to hang out later. Tidy and hoover downstairs and make sure father (in law) is right for the morning. Answer phone three times before 9.00am – all to do with the farm. 'Ministry' arrived at 9.30am to scan 'Chernobyl' sheep – help husband drive and catch these. Finished at 2.30pm – but make lunch for everyone in between and got washing on the line. Went food shopping (15 miles away) and called in at the vet's and farmers' co-op to pick things up for the farm. Did banking and called in at the accountants to sort out some business problems. Home, got dinner for everybody and washing off the line. Husband and son off to (rented) lowland to check stock there, so I fed dogs (5), washed dishes and helped children with homework. Did*

some ironing and went to village for carnival committee meeting (fundraising for community centre). Home around 10.30p.m for supper for family then did some work on accounts before bed at 12.00 midnight.

FIGURE 17.1 Agricultural gender relations portrayed by Britain's toys.

Some women identify strongly as a 'farmer's wife'. For these people, the title recognises the diverse and important role that they play on the farm (as Ashton's extract demonstrates). It also allows them to identify with and support other women in similar roles. Yet the title 'farmer's wife' also suggests that 'women who are married to farmers live their life in the context of their husband's job and that they (and others) define their working life in the context of their marriage rather than the tasks they perform' (Bryant, 1999: 237). One male farmer in Devon said:

> *Well yeah, a good partner as regards husband and wife is, is something that you rely on a hundred-percent. I mean I do . . . I'm in a very, very fortunate position to have a good wife, two sons that can manage to carry on our lifestyle. . . . I can't do any better than that can I?*
>
> *(Kelly and Yarwood, 2018: 102)*

Terms such as 'farmer's son' and 'farmer's daughter' also point to the importance of children to the family farm (Riley, 2009; Robson, 2004), although ideas of 'family farming' often hide or underplay these contributions (Price and Evans 2006). Del Casino (2009: 216) notes that 'the work of child or elder is constituted as non-productive and the appropriation of child labour by parents on the family farm is thought of as obligation and not as economic production'. These labels also help to enrol children into the patrilineal practices necessary for succession in family farming (Leckie, 1996; Price and Evans, 2006). For boys, this can mean operating heavy machinery and learning roles that will enable them to take over the farm (Gasson et al., 1988). For girls, they can be an expectation that they will marry another farmer and fulfil gendered expectations of a 'farmer's wife'. Those who are so are 'good as gold', but those unwilling to conform to these expectations can be treated with suspicion:

> *Ex-wife was a city girl. I should have known she wouldn't take to it. My friends, who married farmers' daughters, they've got on well. They're practical women, turn their hands to anything.*
>
> *('Alun' quoted in Price and Evans [2006: 228])*

The physical labour associated with farming has also led to an association between masculinity and strength (Ni Laoire, 2002). Peter et al. (2000: 226) show that being able to operate 'big machines that control the environment' is an important aspect of masculinity among farmers in Iowa, USA. Likewise, Brandth's (1995) analysis of Norwegian tractor adverts revealed a gendered relationship between technology and farming. She notes that tractors were portrayed as masculine and able to tame and cultivate 'virgin' land. In rural Australia and New Zealand, Liepins (2000a) demonstrates that 'control over land and stock' is central to identity as a 'true' farmer.

The association between physical toughness, masculinity and farming presupposes that male farmers should be mentally tough: 'a masculine hegemony that lauds stoicism in the face of adversity' (Alston, 2012: 515). Evidence is emerging that this contributes to higher risk-taking, a greater chance of occupational accidents, higher rates of suicide and poor mental health amongst men (Shortall, 2016). These situations are exacerbated by a greater reluctance by men to seek help, a lack of mental health care in rural places and cultures of silence or exclusion (Parr and Philo, 2003; Philo et al., 2003; Kelly and Yarwood, 2018). During periods of farming crisis or change, these issues become particularly significant (Convery et al., 2005).

Gender roles and the expectations associated with them are shaped by changing political, economic and cultural circumstances (Shortall, 2016). Many activities associated with farm diversification are often pioneered and administered by women (Brandth and Haugen, 2007). Morris and Evans (2001) note an example of a woman opening and running a post office on her farm. Although she runs the enterprise, her husband describes himself as the 'postmaster' while she continues with childcare and domestic work. They note that:

> *her labour burden has increased whilst there is little evidence of any increase in her decision-making power within the business, or recognition of these issues in the reporting. The nurture associated with emphasised femininity is evident in the fact that her domestic responsibilities remain unchallenged.*
>
> *(Morris and Evans, 2001: 383)*

Farming has relied on a set of gendered roles that are being challenged through economic restructuring and social change. In some cases, changing gender roles are challenging patrilineality (Charatsari and Černič Istenič, 2016), but, in other cases, gendered differences are being asserted through new working practices. Changing farming practices have meant that entrepreneurial, technological, organisational and managerial skills are becoming as valued as much as practical knowledge or manual labour (Little, 2016). Some commentators have suggested that women are better able to adjust to these changes (Brandth, 2002; Charatsari and Černič Istenič, 2016). In one study of dairy farming in Northern Victoria, Australia, it was found that women were moving away from supportive and administrative roles towards managerial tasks and financial accounting. By contrast, having seen their mothers in this role, young men are now coming to consider these as feminine tasks, preferring to busy themselves, like their fathers, with manual work, irrigation and milking (Coldwell, 2007).

Most research on gender and rural geography has focused on farming, which is perhaps surprising given that relatively few people now work in this sector. As the rural economy has diversified into manufacturing and service activities (Chapter 6), this has impacted gender and employment (Henderson and Hoggart, 2003). The urban-to-rural manufacturing shift, for example, was driven in part by an opportunity to exploit non-unionised labour and, in particular, women on part-time, flexible contracts (Keeble, 1980) (Szekely and Michniak, 2009).

Data from the US Census reveal that women working in rural places have the lowest median incomes of the national workforce (Figure 17.2). This reflects poor access to childcare (Halliday and Little, 2001), transport (Law, 1999), financial services (Fletschner and Kenney, 2014) and health (Panelli et al., 2006) that contribute to a jigsaw of exclusion that prevent women from fully participating in the workplace or society. By contrast, men are more likely to commute to jobs in urban areas that offer greater opportunities and career choices (Little, 2002a). More men also work from home, which reflects that they are more likely to hold professional, flexible and better-paid jobs that allow home-working (Department for Environment, Food and Rural Affairs, 2019).

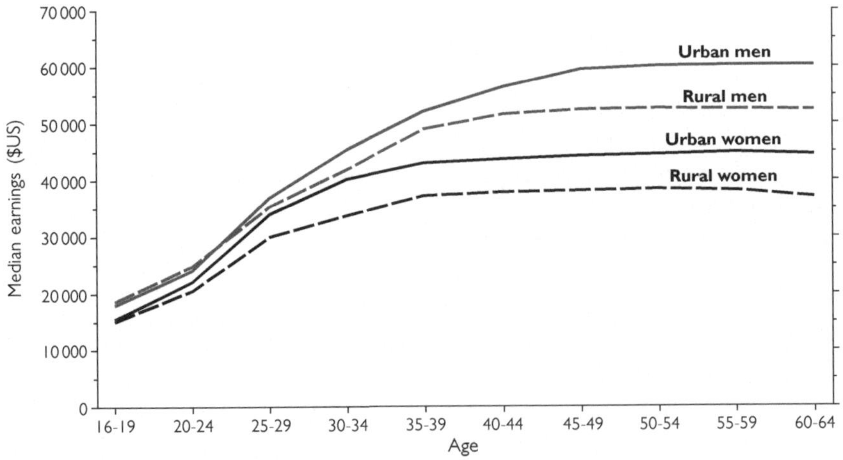

FIGURE 17.2 Median earnings by age and gender in the USA in 2015.
Source: Cheeseman at al 2016

Inequalities in the workplace continue to limit women's opportunities in rural places and, the following section reveals, women can also be further oppressed in domestic and community spaces.

17.3 Domestic Violence, Rurality and Gender

Crime often goes unseen in the countryside due to associations between rurality and safety (Yarwood and Gardner 2000; Mawby and Yarwood 2010). This is particularly true of domestic violence and sexual assault, which are hidden by the private spaces of the home (Wendt, 2009). These offences are notoriously under-reported and under-recorded, making it difficult to compare crime rates in urban and rural areas, but there is some evidence that these crimes are more prevalent in rural places (Dekeseredy, 2015). For example, Owen and Carrington (2015) note 92% of Local Government Areas with the highest rates of domestic violence in New South Wales are 'regional' or 'rural'.

Although domestic violence usually occurs in the home, there is a need to understand it in relation to wider rural society and community (Owen and Carrington, 2015; Little, 2017). Jo Little (2017) has argued there is an association between constructions of gender, rurality and domestic violence. As already noted in this chapter, there is an association between masculinity and *power over* nature or landscape (Brandth, 2006). Little (2017) suggests that this may be extended into the domestic sphere through behaviours that seek to assert dominance and power over other family members: 'men's violence is not "done" to others but is part of their identity, reflecting an important power over rurality rather than a mode of behaviour within rural social relations' (Little 2017: 481). This can normalize violent behaviour and reproduce it across generations. A support worker in Australia commented:

> *I might hear: 'Such and such did this' and: 'The old man was the same'; you know, that sort of thing. . . . Being isolated, they don't see anything else, they just see that. . . . They grow up thinking that is normal.*
>
> *(quoted in Carrington et al, 2013: 8)*

If masculine identity is threatened, perhaps through the loss of work or social status, these tendencies may be further exacerbated.

Increasingly, feminist researchers are conceptualising domestic violence as a form of 'intimate terrorism' (Pain, 2014; Little, 2017). Drawing on the language of geopolitics, they identify that emotional and psychological abuse creates as much, or more, fear as physical violence. These include 'isolation, sexual and economic exploitation, rules about everyday behaviour such as what they should wear and who they should meet' (Little, 2017: 476). This is demonstrated by an interview by Carrington et al. (2013: 7) with a government worker:

> *[two out-of-town women] came up and said: 'I wish I could come and talk to you'; and I said: Oh yes, we could go here or there. But the husbands had forbidden them, and that was the words they used, forbidden them; and that's a form of violence.*

Although domestic violence takes many forms, including physical, emotional and psychological abuse, 'space and place underpin not only the diverse experience of violence but also the behaviours, techniques and social structures through which it is exercised' (Little 2017: 477). Thus, in rural areas, it is an irony that domestic violence, which is so often associated with hidden, private spaces, is supported by wider, public visibility. Hogg and Carrington (2003: 308) explain:

> *in small towns with high levels of mutual recognition, embarrassment acts as a major deterrent to seeking outside assistance. [A woman] . . . could not bring herself to seek outside help or leave her violent husband because of what she imagined members of the community would think of her.*

Similarly, they report a police officer saying:

> *I know of incidents but they don't go to the police because of embarrassment. . . . They don't involve their own town, because they don't want people to know. . . . I think that's a big thing and embarrassment in a small town like this. People know what's going on, and they don't want people to know so they don't turn up.*
>
> *(Hogg and Carrington, 2003: 310)*

The example demonstrates the significance of understanding domestic violence in the social and political context of a wider 'rural architecture' (Owen and Carrington, 2015). Framing domestic violence as a form of terror emphasises the significance of spaces beyond the home; that is political as much as it is personal, reflecting societal structures that continue to allow, or even condone, these forms of intimate violence. Further, as the interviewee implies, rural policing remains focused on public space, and there is an unwillingness to intervene unless somebody comes forward. This is furher exacerbated by sparse policing and a lack of women's refuges in rural places (Coy et al., 2009).

17.4 Governance and Gender

Although new forms of governance are emerging in rural places, they continue to reflect gendered power relations (Little and Jones, 2000). Thus, in one study of a development committee in an Australian shire, Pini (2006) notes that nearly all the board members are men that espoused a 'dominant discourse of managerial masculinity focused on aggression, decisiveness, strength, efficiency, action and vision' (p. 402). Despite a discourse of merit and diversity, Pini argues that existing, male-dominated business and social networks were utilised to appoint members.

Thuesen (2016) suggests that external and internal forms of exclusion contribute to this situation. External exclusion refers to the way in which women may be prevented from participating in governance by existing structures. Drawing on the example of LEADER local action groups, she suggests that women are excluded because of requirements to include members of other committees, such as local governments, that may already be dominated by men. External exclusion may also be used to describe occasions when women exclude themselves if they feel that their experience and skills are not relevant. In Pini's example, cultures of 'managerial masculinity' may leave women feeling unwilling or unable to stand for office.

Internal exclusion refers to the ways that tasks may be ascribed on the basis of gender stereotypes, with women usually taking up 'softer roles'. Thus, Little's (2002b: 146) study of governance in rural development in the UK notes that 'all the chief exec types representing organisations are all men . . . women tend to be involved with community or education'. The idea that certain spaces and activities should be ascribed to particular genders is problematic for two reasons. First, it replicates dominant and sexist views that lead to internal and external exclusion from power. Second, it reaffirms the view that gender is a label that fixes people into certain roles and places. Instead, as the following section examines, gender is an important part of a person's identity that is played out and replicated on a daily basis.

17.5 Gendered Bodies and Rurality

Simone de Beauvoir (1972: 295) argues that gender is not fixed:

> *one is not born but rather becomes a woman. No biological, physiological or economic fate determines the figure that human being presents in society: it is (for example) civilisation as a whole that produces this creative indeterminate between male and eunuch which is described as feminine.*

Gendered identities are constantly being reproduced, challenged and reconfigured through daily lives and actions (Butler, 1993).

Thus, rurality has been equated with embodied forms of masculinity such as:

> *the logger with his chain saw felling a giant tree; the Marlboro cowboy cantering over the plains on his trusty horse; the pioneer leading his wagon across the prairies and defending his family from the howling Indians; the hairy Iron John warrior rising from the swamps to lead modern men out of the crisis of their threatened masculinity; the soldier defending the green fields of the mother country.*
>
> *(Campbell and Bell, 2000: 540)*

In contrast, woman have often been expected to embody ideas of domesticity through tasks such as cooking, baking, mending, sowing or gardening. This was reflected in the foundation of the Women's Institute (WI) in Canada in 1897 with the purpose of teaching women domestic rural skills (Hughes, 1997). The following interview reveals that being a 'country lady' continues to be associated with embodied performances of baking and 'joining in':

> *Some of these country ladies, I don't know how they do it. One particular lady, she cooks, and if there is something going out in the village she will pull her weight and she will bake a few cakes and she will always do her bit to help. She is wonderful cook, a real country women.*
>
> *(quoted in Hughes, 1997: 132)*

These ideas of femininity were challenged by the WI themselves, when the Rylstone and District Branch published a now-famous charity calendar in 2000 (Figure 17.3). The calendar playfully subverted activities usually associated with the Women's Institute – such as baking, jam making and publishing a staid calendar – through the photography of naked women. The calendar raised over

£5 million and became the subject of the feature film *Calendar Girls*. Although women's bodies in the calendar were portrayed as cheeky, homely and 'mumsy' (Little, 2002a: 164), the work was widely feted for celebrating ordinary women's bodies rather than conforming to widely held expectations often shown in the media.

The calendar also demonstrates that bodies are important to gendered identities (Longhurst, 1995). Cloke (2014: 68) outlines:

> *The body orders our access to and mobility in space and places; which interfaces with technology and machinery; which encapsulates our experiences of the world around us; which harnesses unconscious desires, vulnerabilities, alienations and fragmented aspects of self, as well as expressions of sexuality and gender, and which is a site of cultural consumption where choices of food and clothing and jewellery, for example, will inscribe meanings about a person.*

A body is significant in the creation of gendered identities (Little, 2002), it is 'an active and reactive entity which is not just part of us, but is who we are' (Butler, 1999: 239). One example of this is seen in military training. A person becomes a soldier through tough physical training in which masculinity is embodied through an ability to overcome harsh, inhospitable environments:

> *The recruits become exhausted, cold, wet, hungry, and injured, but still they carry on. And through-out, while superior officers urge them on, their identities as men are forged. The sheer physical challenge of route marches and mountain running is presented as a test of one's manhood. The warrior hero must be fit enough to conquer landscapes; indeed, he is literally made in the landscape of the Army's training areas.*
>
> *(Woodward (2000: 651)*

In daily life, gender roles are also being challenged and re-imagined through embodied routines. Some women operate machinery and undertake heavy physical labour that has been associated with men (Brandth, 2006). Despite showing physical toughness, some of these women wanted to maintain emotional qualities associated with femininity: 'to feel powerful and capable with the machines, but not insensitive and hard'. Others have participated in challenging rural sports (Dilley and Scraton, 2010) or drinking contests (Leyshon, 2008b) to challenge the male hegemony of outdoor activities.

These examples highlight that gender is defined, contested and reproduced on a daily basis through embodied practices (Beauvoir, 1972). Until recently, gender was examined as a binary, so that 'the lives and experiences of non-binary and genderqueer people, and other trans folk who live beyond gender binaries are almost non-existent within geography' (Todd, 2021: 7). In rural studies, it was widely assumed that non-binary people moved out of rural spaces to seek the anonymity and diversity of cities, but this is not the case (Abelson, 2016). Research has revealed that some trans-people live hidden lives in rural places, 'passing' as people who were born male for their own well-being and safety (Rogers, 2019). Similarly, a study of trans people living in socially conservative areas of the midwest and southwest of the United States revealed that many trans men emphasised 'sameness' by behaving in ways that could be 'read by others as cisgender, white, and performing appropriate working-class heterosexual rural masculinities' (Abelson, 2016: 1540). Although ideals of rural life motivated some trans people to move to the country, many left their home town to be 'just another guy' in a place where their histories were unknown (Abelson, 2016). This work highlights a need to understand how gendered identities intersect with other forms of identity and rurality (Hopkins and Noble, 2009).

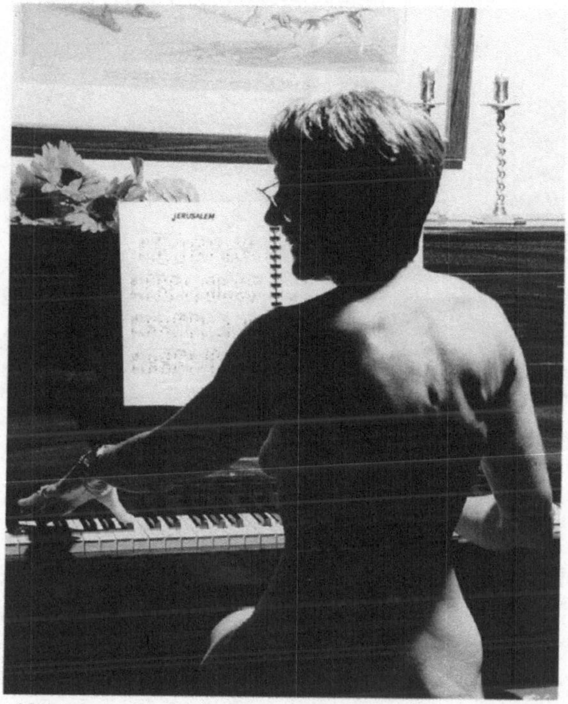

and did the countenance divine shineforth...

February 2000

mon	tue	wed	thurs	fri	sat	sun
	1	2	3	4	5	6
7	8	9	10	11	12	13
14	15	16	17	18	19	20
21	22	23	24	25	26	27
28	29					

FIGURE 17.3 The Rylstone and District WI Branch's Calendar.

Box 17.1 The Naked Rambler

Our bodies, clothes and the spaces that they inhabit are important in the reproduction of cultural norms and what is or is not expected in the countryside. An extreme but nevertheless revealing example of this is illustrated in the case of the Naked Rambler in the UK (Figure 17.4).

FIGURE 17.4 Stephen Gough: The Naked Rambler.
Source: Getty/Jeff J Mitchell/Staff

Stephen Gough is a naturist who has long campaigned for the right to be naked in public. In 2003 he undertook a walk from Land's End to John O'Groats (the length of Great Britain) wearing only boots, socks and his rucksack. He repeated this in 2005, accompanied by his girlfriend, who also walked naked, and a BBC correspondent who filmed the journey. On both occasions, he was variously ignored, celebrated or confronted, sometimes violently, by members of the public. In an interview, he recounts the bodily experiences of engaging with the countryside through naked walking:

> *[I]t's a beautiful sunny day and as long as you've got a lot of Vaseline on your body to stop your rucksack rubbing, you got the sun on you and it's just great, as I wrote to someone once, being naked is like the expression of freedom, being naked is you are not restricted by clothes and the definition of freedom is you are not restricted in any way, it's freedom being expressed in a physical form.*
>
> (Slominski, 2020)

By contrast, in winter months, Gough stated that it was

> *[p]retty cold but manageable . . . when you stop, as that's when you start seizing up. The trick is to keep going . . . When there's snow on the ground, it's hard to*

get out of your sleeping bag, let alone your clothes, to do a 22-mile walk. We wore warm hats, thick socks, gloves and walking boots. We ate lots of carbohydrates and walked fast. The closer we got to the finish, the easier it was to forget the cold and pain.

(Forsyth, 2012)

Gough was arrested twice in England, before being released immediately, and over 18 times in Scotland, where the police took a tougher attitude towards him. In 2006, when onboard a flight to Scotland for trial, he removed his clothes and refused to dress for court. This resulted in a six-month prison sentence. Upon release, he refused to wear clothes, leading to re-arrest, naked court appearances and further imprisonment. This cycle, together with breaching anti-social behaviour orders obliging him to wear clothes, meant that he served ten years in prison, mainly in solitary confinement because of his refusal to wear clothes in prison. Gough unsuccessfully appealed to the European Court of Human Rights that his rights to privacy and freedom of expression had been undermined. Questions have also been raised about his mental health, although Gough insists he walks naked and serves time in prison to campaign for wider freedoms to be naked: 'what I'm doing isn't about me. I'm challenging society and it must be challenged because it's wrong' (Forsyth, 2012).

At the time of writing, Gough is free from prison and caring for his elderly mother. As he would be unable to support her from prison, he is wearing clothes to prevent further arrest but has stated that he will remove them in the future.

Gough's experiences emphasise that walking in the countryside is an embodied and sensory experience that, for most of us, is mediated through various technologies and clothing. Yet Gough's walks have been controversial and demonstrate that the countryside, as well as the ways in which we can behave and respond to it, is controlled by formal and informal sets of practices that encourage certain bodily behaviours and prevent others. His case illustrates the importance of the body and its regulation in determining how to and who can engage with the countryside.

17.6 Sexuality

In 2011, the singer Rihanna filmed a video in a field in Northern Ireland. The landowner had initially given permission for the shoot but withdrew this when her behaviour and clothing became 'inappropriate' in his eyes. Rihanna initially wore a 'demure checked dress' in a grain field (conforming perhaps to a vision of rural innocence) but, according to the horrified farmer, then appeared to be 'in more of a state of undress than a bikini top'. He commented,

I thought it was inappropriate. I requested them to stop and they did. . . . From my point of view, it was my land, I have an ethos and I felt it was inappropriate. . . . I wish no

ill will against Rihanna and her friends. Perhaps they could acquaint themselves with a greater God.

(Singh and Anitasingh, 2011)

As Little (2002a) points out, rurality has strongly been associated with heterosexual, conservative sexualities. Features such as 'Farmer Wants a Wife' and classic heterosexual weddings in *rural* churches or country houses/hotels reinforce an association between rurality, marriage and heterosexual relationships. Women are often expected to be 'coy, demure and rather wholesome' (Little, 2002), and, by contrast, men are expected to conform to a 'rugged masculinity' characterised by outdoor, manual work (Campbell and Bell, 2000).

Bell and Valentine (1995) challenged heteronormative ideals by drawing attention to the lives and representation of gay people in rural areas. They outline a wide set of media representations – from novels to pornography – that present the countryside as a place of sexual but often hidden desire. Annie Proulx's short story 'Brokeback Mountain', for example, tells the story of two cowboys who form an illicit relationship in an isolated mountain setting, which is then hidden from their families and societies on their return. The story, made into a blockbuster film, reflects 'high levels of same-sex sexual activity between rural men precisely because of isolation and conservatism' (Bell and Valentine, 1995: 117).

Bell and Valentine draw attention to the restricted and secretive lives of gay people living in conservative, rural localities, noting problems of 'isolation, unsupportive social environments and a chronic lack of structural services and facilities – leading to eventual or projected emigration to larger (urban) settlements which offer better opportunities for living out the "gay life"' (p. 116). These cultural and social restrictions have led to a complex geography of migration and mobility, often centred on gay people leaving rural places for the anonymity and opportunity offered in urban areas. Drawing on work in the USA and France, Annes and Redlin (2012) argue that, for many gay people, coming to terms with their sexuality and their first sexual encounters occur a city or a college town: 'the city provided a necessary social and political understanding of ... sexual identity but was not a valued place in which to live those newly found ideals' (Annes and Redlin, 2012: 62). Consequently, many people return home or travel between rural home places that are important to their identity. Over time, activism and changing social attitudes have led to a wider acceptance of sexual minorities. Communes (Valentine, 1997a), festivals (Browne, 2011; Gorman-Murray et al., 2012), tourism (Baird, 2006) and gentrification (Smith and Holt, 2005) illustrate how rurality offers a chance for gay people to develop homonormative spaces in the country.

17.7 Conclusions

Geographers are revealing the importance of gender and sexuality in the ways that rurality is imagined and performed. In doing so, they are also developing new ways of doing research that pay greater attention to the researcher's own positionality and role in the research process. Thus, depending on the context of her work and who she was talking to, Pini (2004) variously identified herself as a 'farmers' daughter', 'Italian-Australian', 'nice country girl' and 'woman' when conducting fieldwork in rural Australia. Chiswell and Wheeler (2016) highlight some of the difficulties faced by female researchers working in a masculine, often isolated environments. While safety remains a key concern, they also point out that older male farmers were more likely to reveal their emotions to younger female researchers than to their peers.

Feminist research has also questioned the binary between the researcher and the researched. Very often, the researcher is seen as being more powerful as, for example, they determine what questions are asked in an interview or the focus of a particular investigation. To try and overcome this, feminist researchers have advocated empowering research participants to involve them more fully in designing the research project. In these ways, research aims to allow participants to develop a critical understanding of their own life situation and leads to the development of strategies to challenge inequalities. Pini (2002), for example, pioneered the use of focus groups as a strategy to empower women to produce knowledge that, in turn, can be used to drive social and political change. Studies by feminist geographers are not only contributing to new and significant perspectives on rural space but also to work that can transform that space.

CHAPTER

18

More Than Human Ruralities

18.1 Introduction

For many people, places are only considered rural if they have fields, flowers, trees, wildlife or livestock. But nature and animals are more than simply a backdrop to human affairs; they are agents that shape the country. Livestock, for example, are an integral part of many farms and actively change ecosystems and landscapes. Areas of wilderness are defined by the lack of human activity and the significance of plants and animals in them. Current efforts to rewild places depend on creatures and flora re-establishing *themselves* in places once occupied by people. In all these ways, animals and nature *are* the country.

There has been a long tradition of studying animals in rural and wild places. The study of zoography was concerned with mapping the habitats and density of animal and plant species (H. O. F., 1911; George, 1962; Hesse et al., 1937), and currently, the subdiscipline of biogeography seeks to examine environmental and evolutionary factors that shape the geographical distribution of different organisms (Cox et al., 2016). Yet for a long time, animals and nature seemed to appear in the background of rural geography with no or little acknowledgement of their agency and lives. Chris Philo (1995: 657–658) argued that there was 'something missing: a sense of animals as animals; as beings with their own needs, and (perhaps) self-awareness's, rather than merely as entities to be trapped, counted, mapped and analysed'. For a long time, human geography was, well, too human.

In response to this concern, new 'animal geographies' emerged in an effort to make geography less anthropocentric (Philo and Wilbert, 2000; Urbanik, 2012; Tuan, 1984; Whatmore, 2002; Anderson, 1997; Wolch, 2002; Wolch and Emel, 1995; Jones, 2006; Evans and Yarwood, 1995; Abram et al., 1996; Gillespie and Collard, 2015). The aim was to bring animals 'back in' to geography (Philo, 1995) and recognise their contribution to the country. In addition, outbreaks of diseases such as bovine spongiform encephalopathy (BSE) (Hinchliffe, 2001), foot-and-mouth disease (Convery et al., 2005), tuberculosis (Enticott, 2008) and viruses (Lavau, 2017), including Covid-19, have caused geographers to reappraise our relationship with animals and other non-human agencies. Beginning with farm animals, this chapter outlines some of the ways in which rural geographers have sought to understand the significance of animals and nature in the countryside.

DOI: 10.4324/9780429448966-22

18.2 Farm Animals

Historically, rural geographers concerned themselves with the distribution of different types of livestock in changing agricultural systems (Gilbert et al., 2018). John Fraser Hart's (2003) work revealed how specific forms of livestock farming became spatially concentrated in particular localities of the Americas. This kind of work is important as it reveals the pace and nature of agricultural change, but it does raise questions about how we see and use animals. Animals such as cattle, sheep, pigs, poultry and goats are widely referred to as 'livestock', a word that was first used in the 17th century to describe animals kept or raised on a farm or ranch: stock (goods) that are alive. But this term simply relegates animals to 'units of production' within farming systems (Evans and Yarwood, 1995), as Hart (2003: 40) describes:

> *Cattle have complex digestive systems that enable them to eat grass, leaves, and other parts of plants that many other animals, including the human animal, cannot digest. We depend on cattle to convert these plant parts into food we can eat.*

Our relationship with animals is far more complex than this and reflects cultural as well as economic values. Some animals inspire wonder, others fear or disgust; we want to conserve some, yet eradicate others. We allow some creatures, such as cats and dogs, into our homes as companions, yet others are reared, often out of sight and mind, as food. In turn, what we consider food reflects social, cultural and religious differences. Some religions abstain from pork or revere some animals as sacred. These preferences represent different ways in which animals are socially or culturally 'constructed' by different people.

Some of these relationships can be revealed by examining the distribution breeds, rather than species, of animals in farming systems (Yarwood and Evans, 2000). A breed refers to animals of the same species that share a common appearance, characteristics and genetic material, such as Friesian cattle, Old Spot pigs or Rhode Island Red chickens (Evans and Yarwood, 1995). What constitutes a breed is determined by breed societies, which can also reflect particular agricultural and societal interests (Yarwood and Evans, 2006). Breeds have distinctive geographical distributions that reflect strong historical and cultural associations between breed and place (Yarwood and Evans, 1999; Evans and Yarwood, 1998; Walton, 1984).

Some breeds are clustered in their areas of origin (Yarwood and Evans, 2006). For example, North Devon (or Red Ruby) cattle are still mainly found in North Devon and South Devon cattle in South Devon (Yarwood and Absalom, 2006). This is in part explained by the characteristics of these breeds, which are particularly suited to grazing the landscapes of these localities because of decades of selective breeding, but this does not explain why these breeds are chosen over more commercially viable animals nor why these breeds are absent from other parts of the UK with similar environments to Devon.

Their concentration reflects a strong association between breed, place and local farming practices that makes farmers more likely to farm these animals. These traditions are strengthened through the prevalence of small and medium family farms in Devon. The attachments between place and breed are also evident in a marketing campaign for Holsworthy in North Devon, which promoted itself as 'Ruby Red

Country' in recognition of the distinctive North Devon cattle found there. South Devon cattle are referred to affectionately as 'Orange Elephants', due to their size and colour, and have been used in the marketing of local produce.

Rare, local or unusual breeds have also been re-purposed as visitor attractions in farm parks. In these cases, animals are valued for their appearance and peculiarity (Figure 18.3), rather than their produce. As such, the revival of some breeds reflects a trend of farm diversification (Chapter 5)

By contrast, more commercial breeds are distributed more widely (Tonts et al., 2010; Yarwood et al., 2010; Yarwood and Evans, 2003). The Lleyn sheep originated in Gwynedd, North Wales, and is now found across the United Kingdom and the world (Yarwood and Evans, 2006). Originally the remoteness of the Lleyn peninsular meant that the sheep attracted little interest from farmers outside the locality. This was so much so that at one point in the 1970s, only ten flocks remained, all in North Wales. The number and geographical distribution of the breed increased dramatically when it was crossed with other breeds to produce high-quality meat.

FIGURE 18.1 Unusual animals – such as this five-legged sheep in New Zealand – are important for visitor attractions and underline how animals are socially and physically constructed by people.

The geographies of livestock breeds point to some of the complex relationships people have with animals. To a large extent, this reflects the ways in which animals are socially constructed and physically manipulated by people to fit into certain spaces. Thus, Buller and Morris (2003) describe livestock as 'quintessential hybrids' because, while we often deny their existence as animals we also value their apparent naturalness. In response to this dilemma, some studies have examined how the emotional and affective connections between farmers and animals extend beyond economic or functional values (Wilkie, 2005). Lewis Holloway (2001), for example, examined the ethical dilemmas faced by small-holders when taking their animals to slaughter, questioning whether they regarded them primarily as pets or food. One farmer reported taking a goat to slaughter: 'I drove into a lay-by and wept buckets – because I'd bottle-fed him' (quoted in Holloway, 2001: 304).

This is not confined to hobby farmers. During the 2001 foot-and-mouth disease (FMD) outbreak in the UK, many livestock were slaughtered to prevent the disease from spreading. This was traumatic for many farmers who witnessed the destruction of animals they knew intimately:

> *[T]hey all had names here, they weren't just numbers, they all had names and known individually. They are all characters and individuals in a way when you farm the way we do.*
> (*Farmer quoted in Convery et al., 2005: 105*)

A striking set of photographs by Chris Chapman titled *Silence at Ramscliffe* (Figure 18.2a and b) traced the final milking of a herd before they were slaughtered, leaving silence and empty fields (Yarwood, 2005a; Chapman and Crowden, 2005). The photographer commented:

> *I have often wondered why so many war photographs are technically poor. The images often suffer from too much contrast or are badly focused. The end result is simply raw. Let me tell you that using a camera in such circumstances is no easy task. I am used to photographing everyday life, its ups as well as downs, but this was surreal and I pressed the shutter all day long as if in a dream.*
> (*Chapman, 2004*)

FIGURE 18.2A Silence at Ramscliffe: MAFF employee monitoring the sedation of the dairy herd, Ramscliffe, Beaford, North Devon, © Chris Chapman 2001.

FIGURE 18.2B Silence at Ramscliffe: Pithing then spraying the carcasses with dye, Ramscliffe, Beaford, North Devon, © Chris Chapman 2001.

Chapman's photographs were also included in Devon's public inquiry into the disease in order to convey some of the emotional stress faced by farmers at the time (Mercer, 2002).

In light of these complex but often hidden relationships, geographers have attempted to pay more attention to the ways in which human and animal lives are interlinked in different places. Convery et al.'s (2005) analysis of FMD led to them identifying how time (such as the lifespan of an animal and the intensity of daily contact), scale (the type of farming) and location (distant or close to the farm) frame the relationship between farmers and animals. These range from close associations – 'dairy cattle are different aren't they? People don't realise that, they're not cows they're your friends . . . my dairy cows are my friends' (farmer quoted in Convery et al, 2005: 6) – to clinical appraisals of their value: 'Good functional rumps, clean boned leg with steep foot angle results in cattle that move easily and have trouble free feet and legs. Udders are well attached, high and wide. Excellent teat size and placement makes for easy milking' (breed catalogue quoted in Convery et al, 2005: 6).

In recent years, geographers have moved away from seeing cattle as 'livestock' that are transformed into 'commodities' and, instead, have considered how they play active roles in making places and landscapes (Evans and Yarwood, 1995; Sellick and Yarwood, 2013). Yet viewing animals simply as cultural constructions continues to deny a sense of animals as animals (Philo, 1995) or, as Murdoch (2003: 264) puts it, that 'there is something beyond "the social" at work as the countryside displays a material complexity that is not easily reducible to even the most nuanced social categories'. Animals are sentient beings that shape places with their own lives and habits:

> *Farm animals . . . will always be largely constructed and confined by their human-serving functionality . . . as the property of human individuals, collectives and organisations, their geography and spatiality will – to a greater or lesser extent – be intricately linked to our own. And yet . . . we do not hold total sovereignty over them . . . even farm animals remain, for all their*

breeding, selection, docility, and husbandry, beyond our complete societal appropriation (Buller and Morris, 2003: 217).

This is seen in the example of hefting, the term given to describe the relationship between a flock of sheep and a particular piece of unfenced land (Gray, 1996, 2014). Initially, sheep are shepherded, but over time, they associate with this land to such an extent that, even when untended, they will only graze there. This 'hefting knowledge' is passed from ewe to lamb over succeeding generations of the same flock. This is so much so that there is a 'landlord's flock' that remains with the farm whenever there is a change of tenant.

18.3 Rewilding

Hefting shows the importance of an animal's agency in shaping landscape, but it is important to remember that it is people who have largely determined where animals can live and how they can behave there (Wolch and Emel, 1998; Buller, 2004). This is illustrated when wild animals are trapped, removed or shot if they threaten human settlements or activities (Collard, 2012). More significantly, human habitation and cultivation have reduced biodiversity and contributed to environmental damage (Monbiot, 2014).

In an effort to reverse these trends, the concept of rewilding is becoming important. Rewilding is an attempt to return land to uncultivated, unmanaged state (Jørgensen, 2015; Brown et al., 2011). It originated in the Wildlands Project in the USA that 'aimed to create core wilderness areas without human activity that would be connected by corridors' (Jørgensen, 2015: 483).

Rewilding has been characterised by three Cs: cores, corridors and carnivores (Soule and Noss, 1998). Cores are protected areas of a habitat that, (un)managed by people, return to a wild state; corridors connect these places together to allow the migration of animals; and carnivores refer to the reintroduction of large predators, such as wolves (Buller, 2004), to regulate the food chain. Other animals, including beavers (Law et al., 2017), are also valued for their contribution to maintaining ecosystems or habitats. The presence or absence of certain animals reflects wider debate about what 'wild' means; the extent to which different places are or should be given over to the 'wild'; and over what time periods rewilding should occur (Table 18.1, Figure 18.3) (Deary and Warren, 2017).

TABLE 18.1 Types and Spaces of Rewilding

Summary of Reference Points in Rewilding Definitions		
Rewilding Definition	**Reference Time**	**Geography**
Cores, corridors, carnivores	Up to 4000 BP, but most are within the last 200 years	North America
Pleistocene megafauna replacement	13,000 BP	North America
Island taxon replacement	16th–19th century, depending on a specific island	Islands
Landscape through species reintroduction	Before species extirpation	Europe
Production land abandonment	Up to the Neolithic (c. 6000 BP)	Europe
Releasing captive-bred animals to wild	When captive population created	Any

Source: Jørgensen (2015)

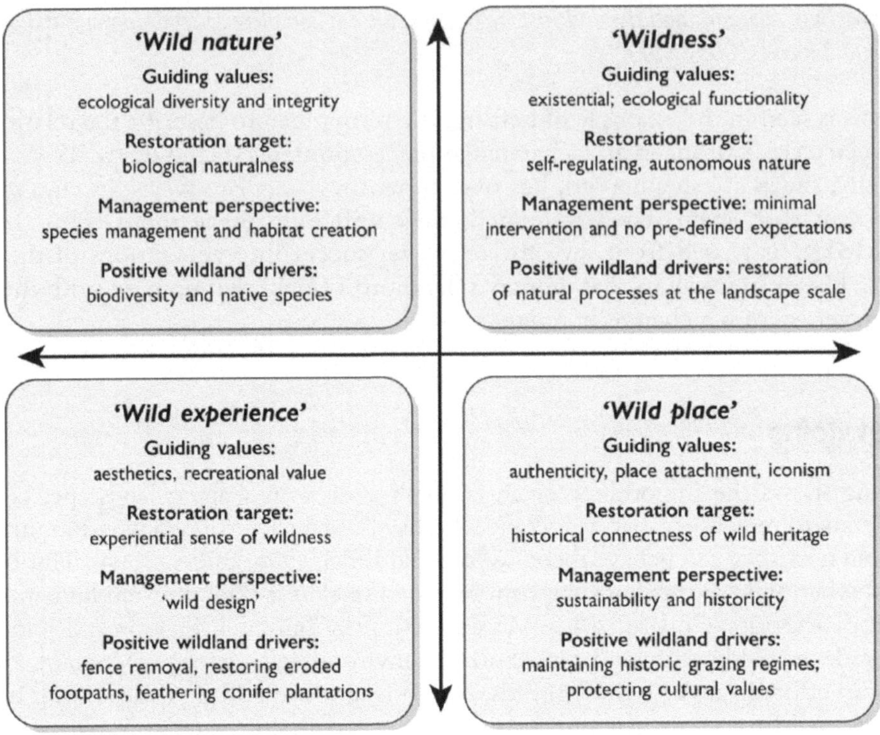

FIGURE 18.3 Forms of rewilding.
Source: Deary and Warren (2017)

The concept of rewilding has raised questions about which animals and fauna belong in particular places. Memorably, the environmental campaigner George Monbiot (2013) railed against British upland landscapes that have been 'shagged by the white plague':

> *Sheep have reduced most of our uplands to bowling greens with contours. Only the merest remnants of life persist. Spend two hours sitting in a bushy suburban garden and you are likely to see more birds and of a greater range of species than in walking five miles across almost any part of the British uplands. The land has been sheepwrecked.*
>
> *(Monbiot, 2013)*

Instead, he argues, there should be a return to the climax landscapes of forested hilltops:

> *Foreigners I meet are often flabbergasted by the state of our national parks. They see the sheep-wrecked deserts and grousetrashed moors and ask: 'What are you protecting here?' In the name of 'cultural heritage' we allow harsh commercial interests, embedded in the modern economy but dependent on public money, to complete the kind of ecological cleansing we lament in the Amazon. Sheep farming has done for our rainforests what cattle ranching is doing to Brazil's. Then we glorify these monocultures – the scoured, treeless hills – as 'wild' and 'unspoilt'.*
>
> *(Monbiot, 2009)*

Monbiot makes a case for the environmental and social benefits of rewilding (Monbiot, 2014), yet his vision runs counter to the cultural and traditional significance of sheep farming in particular localities (Gray, 2014; Squire, 1993), as well as the clear, picturesque views that many have come to expect from hilltops (Macnaghten and Urry, 1998). These debates remind us that animals and plants are socially constructed; the extent to which they seem in or out of place reflects different and often competing ways in which people are culturally conditioned to see animals. People, society and culture determine where and how rewilding occurs.

Yet viewing animals simply as cultural constructions continues to deny a sense of animals as animals (Philo, 1995) or, as Murdoch (2003: 264) puts it, that 'there is something beyond "the social" at work as the countryside displays a material complexity that is not easily reducible to even the most nuanced social categories'.

18.4 Hybrid Networks

We often talk about the countryside in binary terms: nature and society, people and animals, the country and the city, and so on. Yet new animal geographies have challenged the assertion that there are clear-cut differences between humans and non-humans. Instead, as Murdoch (2003: 264) suggested, the countryside is 'co-constructed by humans and non-humans, bound together in complex inter-relationships'. Instead of seeing the country's constitute parts separately, there is an appreciation that they are bound together into complex networks or assemblages (Whatmore, 1997).

One example of this can be seen in the way that specially trained dogs are part of wider human-animal networks that come together to search for missing people (Figure 18.4) (Yarwood, 2015). In these situations, people and dogs work together to locate a casualty through the hound's sense of smell. Key to the success of these searches is for a dog's handler to appreciate how a dog is able to smell and the impact of terrain and wind on scent patterns:

> *Just remember, you're not searching the ground, you're searching the wind that blows over the ground. So wind direction is absolutely key to us when you are searching.*
>
> *(quoted in Yarwood, 2015: 7)*

In the same way, a farmer also uses a dog to round up a flock of sheep on a particular heft. Both the farmer and the dog are essential to this action and co-construct how the sheep are managed. Seen thus, farms are 'collections of things which must be held together, or positioned in relation to each other, so that something like farming becomes an ordering of land, animals, people, etc. which produces the "farm"' (Holloway, 2002: 2057). The farm is a collection of different entities – human and non human – that are enrolled together to form the space of the farm, which is, in turn, connected by other networks to other places (Murdoch, 2000). A dairy farm, for example, would be nothing without cows, but equally, human technologies have been used for breeding cows in order to produce milk (Holloway et al., 2014) that draw on databases of genetic information (Holloway et al., 2009). Robotic milking parlours not only milk cows but also monitor their yield and link these data to national and international distribution chains (Shortall et al., 2018) (Holloway et al 2014).

These 'actor network' approaches are helpful because they stress the inter-connected and co-constructed nature of the countryside. They have extended beyond the study of animals to encompass other 'natural' features such as rivers (Kortelainen, 1999), trees (Jones and Cloke, 2002) and weather (Ingold, 2010). For example, attention has been given to the ways in which nature has been enrolled, rather than controlled, to create orchards (Cloke and Jones, 2001), burial sites (Yarwood et al., 2015) and gardens (Power, 2005).

The intention is to reduce or 'flatten' hierarchies that place society above nature and, instead, emphasise the essential contribution of non-human agencies to rural networks (Woods, 1998). For some, this is a way of empowering non-humans and enrolling them into political decision making. Thus, Gareth Enticott (2001) argues that the 'voice of the badger' should be heard in discussions about whether culling to prevent the spread of bovine tuberculosis (TB). By this, he contends that badgers 'subvert those categories and identities humans place on them' (Enticott (2001: 162) by, for example, recolonising territories that had been culled and, in doing so, threatening the validity of statistical trials.

Ultimately, though, this depends on how badgers are considered by people in these debates. Badgers have been viewed as 'nature as numbers' (TB statistics), 'nature as known' (how nature is known by farmers and other human actors) and as 'ecological nature' (the actions of animals themselves). In the same way, representations of animals have been enrolled into arguments about hunting, where foxes have been presented as both pests and victims (Woods, 1998). While recognising

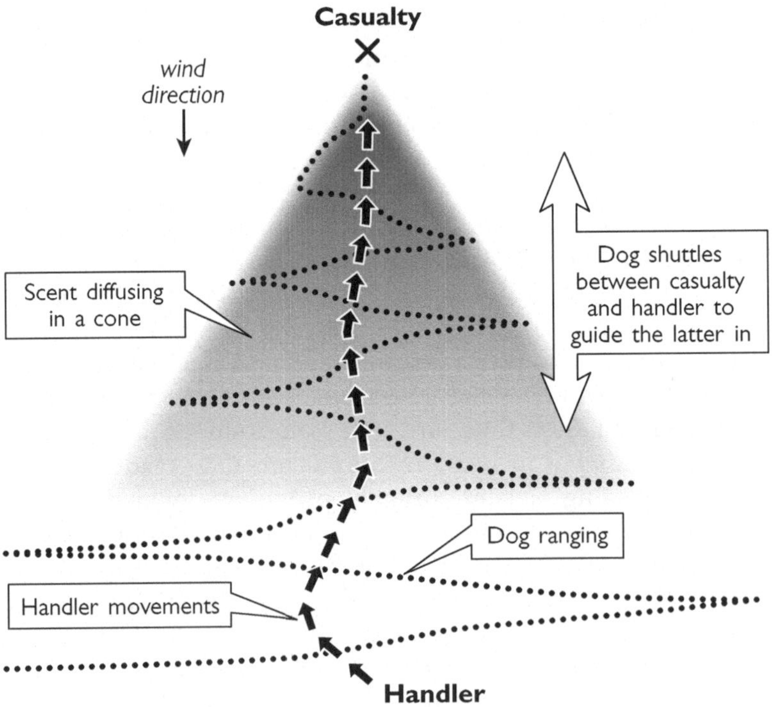

FIGURE 18.4 The relationship between a search dog and humans.
Source: Yarwood (2015)

the agency of animals, it remains important to consider how different groups of people position animals within decision-making frameworks and the extent to which different actors (both human and non-human) have the power to affect the outcomes. As Murdoch (2003: 279) summises 'the countryside continues to be shaped by social processes, it is just that these processes are *more than* social (just as the countryside continues to be shaped by processes that are *less than* natural).'

18.5 Conclusions: Ethical Animal Geographies

Animal geography tries to appreciate that the world is 'more than human' and that non-humans have their own agency, actively co-making and changing places with other non-human and human actors. As the bodies of farm animals are inscribed with antibiotics, pasture pesticides, antioxidants, flavour enhancers, salmonella, growth hormones, antibacterial drugs, sheep offal, listeria, preservatives and meat colourings (Whatmore, 1997), they have become cyborgs (Haraway, 2013) to be exploited by people in various spaces and places (Sellick, 2020). Questions are increasingly being asked about the morality of treating animals in this way, and so perhaps the most important role of animal geography is to interrogate people's ethical relationships with animals and how these shape rural places.

CHAPTER

19 Conclusions

19.1 The Geographies of Rural Futures

While change has been a constant theme in rural geography, most work has examined how the countryside has changed or is changing rather than how it *will* change (Lowe and Ward, 2009). Shucksmith (2018: 165) comments that 'referring to the future and the rural in the same breath may appear to be something of an oxymoron'. Despite this 'present-tense focus' (Woods, 2012: 125), rural geographers have been obliged to think more carefully about the future of the countryside, especially in light of the challenges faced by climate change and the growing need to live sustainably. Three themes are emerging.

First, there is growing concern about the environmental impact of some rural activities and lifestyles. Although urban areas are often associated with unsustainable living and high carbon footprints, it has been mooted that many aspects of rural life are just as or more damaging to the environment. For example, the reliance on fossil fuels for farming and commuting, the environmental damage caused by intensive production and the destruction of wilderness areas are all causes for concern. Further, measures to conserve or preserve areas of open countryside contribute to a 'sustainability trap' (Taylor, 2008), in which restrictive planning policies have led to a preserved countryside that is attractive (and perhaps planned for) middle-class migrants.

Second, certain trends are growing in significance and impact. The rewilding movement is gaining momentum and challenging whether land should be cultivated or managed for people. Equally, a growing trend for plant-based diets is also raising questions about the ethics of farming and human-animal relations. While confrontations can make headlines, changing consumer habits and farming practices are incremental but significant changes.

Although rural areas are often portrayed as being affected by exogenous change, they also offer the possibility of radical, far-reaching change from within. The spatial evolution of society often starts in peripheral rather than core areas (Dodgshon, 1987), pointing to the possibility of rural change driving broader

DOI: 10.4324/9780429448966-23

societal changes. Rural places can offer space for new, more radical forms of citizenship to emerge. The imagined and literal edges of rurality (Halfacree, 2007) have provided spaces for new utopian communities that are based on faith, gender, green politics, political extremism, nomadism or a desire to live sustainably (Spanier, 2021).

Indeed, the connections wrought by a global countryside offer the possibility of radical and transnational politics and resistance emerging from *within* rural places (Woods, 2016). In South America, peasant movements have successfully mobilised indigenous identities to address common concerns. Building on social networks left in place by prior rounds of political and religious organising, indigenous groups have used unions, churches, nongovernmental organisations and even state networks to mobilise across communities to demand rights and resources (Yashar, 1998). One example is the Via Campesina organisation that was formed in Belgium in 1993 to defend small-scale agriculture against corporate and transnational companies. It aims to bring together 'peasants, small and medium-size farmers, landless people, women farmers, indigenous people, migrants and agricultural workers from around the world' and claims to have 164 organisations in 73 countries representing 200 million farmers.

Third, if these movements point to the possibility of change, there is also evidence that a certain status quo remains in rural places.

Thirty years ago, Cloke (1993) noted gloomily that rural problems tend to be repeated with new cohorts of people. This book has highlighted that many of these problems continue and have perhaps got worse. While new residents have brought new concerns and a 'new politics of the rural' that, to an extent, old class structures power remains in the hands of elite groups, albeit different ones. For example, while policy has moved from top-down to bottom-up ways of decision-making and localism is becoming more significant, there is evidence that community groups are still dominated by articulate middle-class groups with their own vested interests in mind. This is one reason why it is so difficult to build affordable housing. More pessimistically, little seems likely to change without radical intervention.

The Covid-19 pandemic has pointed to the importance of rural places and, in particular, the globalised countryside in the supply of food and goods. In the West, it was shocking to find supermarket shelves empty of basic products during the pandemic. As international travel became restricted, concerns were raised that supply chains would be disrupted and that there would not be enough migrant workers to pick food. In response to what was seen as a national crisis, some urban residents took to working in agriculture for the first time and special flights of seasonal workers were arranged to maintain production. In the UK, further concerns have been raised about food supplies during the transition from the EU with, at one point, lengthy queues of trucks forming on either side of the Channel. These are perhaps reminders of what much of the world faces on a daily basis. As I write this, 30 million people in 20 countries, including South Sudan, North Korea, Ethiopia, Afghanistan and Burkina Faso, are facing starvation.

Rural futures are, therefore, complex. While it is possible to advocate visions of future sustainability, there is also a need to understand economic, social and cultural concerns and, significantly, how these are played out in particular places and spaces. Change, as this book has shown, is geographical, and there are likely to be spaces of intense change as well as those that are 'passed by'. Rural geography allows us to understand this complexity.

19.2 The Future of Rural Geographies

Geography is crucial to understanding the changing countryside. Rural geography, as this book has emphasised, is not a stand-alone subject but, rather, part of the wider discipline of geography. As such, it has benefitted from a rich history of academic thought that has sought to understand people's relationship with the environment and the significance of space and place to the organisation of society (Couper, 2014). Geography straddles the humanities, social science and sciences in ways that are unique to the discipline (Jenkins, 2007).

While geography is able to provide a holistic vision of the world, a holistic vision of geography has been elusive and, in any case, not desirable (Castree, 2012). Rather, the strength of geography is in its philosophical, methodological and empirical diversity. It is able to draw on a cornucopia of ideas, traditions and approaches to make sense of the world in a manner that shows awareness of society and the environment. As a discipline, geography should provide visions of the world that enrol people and nature, but at the same time, it should not be proscriptive as to how this should be done.

At times, rural geography has shied away from the theoretical ideas that have been central to geography (Cloke, 1989, 2006a) to such an extent it has been called a 'Cinderella subject' (Cloke, 1980).

Yet at other times, rural geography has been central to the development of geographical thought. For example, during the early 20th century, the concept of the pay or region had a largely rural focus. During the post-war quantitative revolution, Von Thünen's (1966) 'isolated state' model strongly influenced the teaching of geography in universities and schools, as demonstrated by prominent textbooks (Chisholm, 1962; Bradford and Kent, 1993), and so influenced geography for a considerable time (Block and DuPuis, 2001). Rural geography was particularly influential during the 1990s (Cloke, 1997), propelled by the 'excitements' of the cultural turn and, in particular, Philo's (1992) intervention about rural others. It became more theoretically informed and opened up new avenues of research that stretched outside the country to influence the discipline of geography as a whole. Animal geographies and relational geographies, for example, were strongly influenced by rural themes, such as livestock, animal diseases or relationships with nature and wilderness. Recently, rural geography has developed a more global perspective (Woods, 2007) that has not only provided new viewpoints for rural geography but has led to new understandings of globalisation.

So going forward, rural geography remains in rude health. While some trends and issues are emphasised at particular times, the strength of the discipline is its breadth. Rural geography embraces the empirical mapping of crop distribution; the practices of ploughing the fields that grow the crop; and how people, technology and nature that produce food are incorporated into global networks. This diversity is rural geography's asset. Rural geographers have at their command a plethora of approaches that can be drawn upon, as appropriate, to provide insights into different aspects of the country. Geographers should, of course, clasp new ideas but, at the same time, be alert to continuing possibilities of understanding that are offered by established ideas and approaches. To maintain its vibrancy, rural geography should be:

1 Critical: rural geography should continue to be driven by critical thinking in order to understand and analyse the complexities of the country. In the main, this will be strongly but not exclusively influenced by wider theories, paradigms and methodologies in geography and other social sciences. With increasing emphasis placed on inter- and cross-disciplinary studies, even wider possibilities emerge here, but at the same time, the importance of spatiality should continue to be emphasised. We should also be mindful of Cloke's (2006) appeal that rural geography should both reflect and influence the direction of the wider discipline.

2 Relevant: poverty, environmental damage, repression of women and lack of opportunity are apparent in the country. The UN's Sustainable Development Goals, which aim to tackle these concerns and others, therefore demand a rural focus. As such, rural geography should contribute to addressing these challenges.

3 Applied: Rural geography has a long association with policy, practice and planning. Agricultural geography, for example, has often been concerned with analysing and critiquing a maze of policy directions and initiatives. This empirical focus is a core part of rural geography and should be retained. This does not mean, however, that rural geography should be empiricist. Instead, theories and ideas from rural geography should be applied to rural places, be it by contributing to policy and policy analysis or shaping planning policy. As Cloke (2006: 26) argues, the cultural turn should not be divorced from this but, rather, has 'the potential to add significantly to the broader understanding of, and critical importance of, rural policy agendas'.

4 Pedagogic: one of the main strengths of geography is its educational value. While this book has emphasised research, it has been written to educate. Teaching rural geography and encouraging new generations of geographers to think and act critically about rural places is crucial if the subject is to maintain its vibrancy and relevance.

5 Global: rural geography, rather like the communities it studies, has been rather insular. Its literature has been by, from and about the Global North (something that is reflected in the focus of this book). During the period in which this book was written, the Black Lives Matter movement has prompted reflection about the way knowledge is produced. There have been attempts to decolonise geography by destabilising established binaries and empowering those who are seldom given a voice. Although rural geography is becoming more global in its scope, future work should be by, rather than of or about, those from the Global South.

Future rural geographies will continue to be shaped by the wider discipline of geography. As such, they will continue to make important contributions to the ways the countryside is understood, as well the study of geography. Rural geography will continue to build on metaphorical fields of rural knowledge, and just as a developer builds on real fields, this will provoke conflict, objections to change and a yearning for a lost age. So, as rural geography discovers new, exciting pastures, it is important to remember the academic traditions and histories that continue to shape it.

Bibliography

Abelson M. (2016) 'You aren't from around here': Race, masculinity, and rural transgender men. *Gender, Place & Culture* 23: 1535–1546.

Abid M, Schneider U and Scheffran J. (2016) Adaptation to climate change and its impacts on food productivity and crop income: Perspectives of farmers in rural Pakistan. *Journal of Rural Studies* 47: 254–266.

Abram S, Murdoch J and Marsden T. (1996) The social construction of 'Middle England': The politics of participation in forward planning. *Journal of Rural Studies* 12: 353–364.

Ackrill R. (2000) *Common Agricultural Policy*. Sheffield: Sheffield Academic Press.

Ackrill R. (2008) The CAP and its reform – half a century of change? Die GAP und ihre Reform – 50 Jahre im Wandel? La PAC et sa réforme – Un demi-siècle de changement? *EuroChoices* 7: 13–21.

Adey P. (2017) *Mobility*. London; New York: Routledge.

Adger W. (2000) Social and ecological resilience: Are they related? *Progress in Human Geography* 24: 347–364.

Agyeman J and Sooner R. (1997) Ethnicity and the rural. In: Cloke P and Little J (eds) *Contested Countryside Cultures: Otherness, Marginalisation, and Rurality*. London: Routledge, 197–217.

Allamby L, Bell J, Hamilton J, Hansson U, Jarman N, Potter M and Toma S. (2011) *Forced Labour in Northern Ireland: Exploiting Vulnerability*. York: Joseph Rowntree Foundation.

Alston M. (2000) Rural poverty. *Australian Social Work* 53: 29–34.

Alston M. (2002) Social capital in rural Australia. *Rural Society* 12: 93–104.

Alston M. (2012) Rural male suicide in Australia. *Social Science & Medicine* 74: 515–522.

Amin A. (1996) Beyond associative democracy. *New Political Economy* 1: 309–333.

Anderson K. (1997) A walk on the wild side: A critical geography of domestication. *Progress in Human Geography* 21: 463–485.

Anderson J and Peters K. (eds) (2016) *Water Worlds: Human Geographies of the Ocean*. London: Routledge.

Andersson A, Höjgård S and Rabinowicz E. (2017) Evaluation of results and adaptation of EU rural development programmes. *Land Use Policy* 67: 298–314.

Andersson M, Lavesson N and Niedomysl T. (2018) Rural to urban long-distance commuting in Sweden: Trends, characteristics and pathways. *Journal of Rural Studies* 59: 67–77.

Annes A and Redlin M. (2012) Coming out and coming back: Rural gay migration and the city. *Journal of Rural Studies* 28: 56–68.

An Oifig Buiséid Pharlaiminteach/Parliamentary Budget Office. (2018) *An Overview of the Common Agricultural Policy (CAP) in Ireland and Potential Regional and Sectoral Implications of Future Reforms*. Dublin: Tithe an Oireachtais/House of Oireachtas.

Arcury T, Gesler W, Preisser J, Sherman J, Spencer J and Perin J. (2005) The effects of geography and spatial behavior on health care utilization among the residents of a rural region. *Health Services Research* 40: 135–156.

Arensburg C and Kimball S. (1940) *Family and Community in Ireland.* Cambridge, MA: Harvard University Press.

Argent N. (1999) Inside the black box: Dimensions of gender, generation and scale in the Australian rural restructuring process. *Journal of Rural Studies* 15: 1–15.

Argent N. (2002) From pillar to post? In search of the post-productivist countryside in Australia. *Australian Geographer* 33: 97–114.

Argent N. (2017a) Rural geography I: Resource peripheries and the creation of new global commodity chains. *Progress in Human Geography* 41: 803–812.

Argent N. (2017b) Rural geography II: Scalar and social constructionist perspectives on climate change adaptation and rural resilience. *Progress in Human Geography* 43: 183–191.

Argent N. (2018) Heading down to the local? Australian rural development and the evolving spatiality of the craft beer sector. *Journal of Rural Studies* 61: 84–99.

Argent N, Smailes P and Griffin T. (2007) The amenity complex: Towards a framework for analysing and predicting the emergence of a multifunctional countryside in Australia. *Geographical Research* 45: 217–232.

Argent N, Tonts M, Jones R, and Holmes J. (2013) A creativity-led rural renaissance? Amenity-led migration, the creative turn and the uneven development of rural Australia. *Applied Geography* 44: 88–98.

Argent N, Tonts M, Jones R, and Holmes J. (2014) The amenity principle, internal migration, and rural development in Australia. *Annals of the Association of American Geographers* 104: 305–318.

Arnell N and Gosling S. (2016) The impacts of climate change on river flood risk at the global scale. *Climatic Change* 134: 387–401.

Arnstein S. (1969) A ladder of citizen participation. *Journal of the American Institute of Planners* 35: 216–224.

Ashmore F, Farrington J and Skerratt S. (2015) Superfast broadband and rural community resilience: Examining the rural need for speed. *Scottish Geographical Journal* 131: 265–278.

Ashton S. (1994) The farmer needs a wife: Farm women in Wales. In: Aaron J, Rees T, Betts S, and Vincentelli M. (eds) *Our Sisters' Land: The Changing Identities of Women in Wales.* Cardiff: Cardiff University Press, 122–139.

Ashworth G and Kavaratzis M. (2015) Rethinking place branding. In: Ashworth G and Kavaratzis M (eds) *Rethinking Place Branding.* New York: Springer.

Asian Disaster Reduction Center. (2020) *Natural Disaster Data Book 2020 (An Analytical Overview).* Japan: Asian Disaster Reduction Centre.

Askins K. (2006) New countryside? New country: Visible communities in the English national parks. In: Neal S and Agyeman J (eds) *The New Countryside? Ethnicity, Nation and Exclusion in Contemporary Rural Britain.* Bristol: Policy Press, 149–172.

Askins K. (2009) Crossing divides: Ethnicity and rurality. *Journal of Rural Studies* 25: 365–375.

Atterton J. (2016) Invigorating the new rural economy. In: Shuksmith M and Brown D (eds) *Routledge International Handbook of Rural Studies.* Oxford: Routledge; London: Routledge, 165–180.

Aust R and Simmons J. (2002) *Rural Crime: England and Wales Home Office, Statistical Bulletin.* London: Home Office.

Austin E, Rich J, Kiem A, Handley T, Perkins D and Kelly B. (2020) Concerns about climate change among rural residents in Australia. *Journal of Rural Studies* 75: 98–109.

Aygeman J and Spooner R. (1997) Ethnicity and the rural environment. In: Cloke P and Little J (eds) *Contested Countryside Cultures.* London: Routledge, 197–217.

Bailey I, Hopkins R and Wilson G. (2010) Some things old, some things new: The spatial representations and politics of change of the peak oil relocalisation movement. *Geoforum* 41: 595–605.

Baird B. (2006) Sexual citizenship in 'the New Tasmania'. *Political Geography* 25: 964–987.

Baker C. (2015) Beyond the island story? The opening ceremony of the London 2012 Olympic Games as public history. *Rethinking History* 19: 409–428.

Barcus H and Halfacree K. (2017) *An Introduction to Population Geographies: Lives across Space*. London; New York: Routledge, Taylor & Francis Group.

Bardsley D, Weber D, Robinson G, Moskwa E and Bardsley A. (2015) Wildfire risk, biodiversity and peri-urban planning in the Mt Lofty Ranges, South Australia. *Applied Geography* 63: 155–165.

Barnes, T. (2015). "Desk Killers": Walter Christaller, Central Place Theory, and the Nazis. In: Meusburger, P., Gregory, D., Suarsana, L. (eds) *Geographies of Knowledge and Power. Knowledge and Space*, Springer, Dordrecht. pp187-201.

Barratt P. (2012) 'My magic cam': A more-than-representational account of the climbing assemblage. *Area* 44: 46–53.

Barrett H, Storey D and Yarwood R. (2001) From market place to marketing place: Retail change in small country towns – The changing urban hierarchy. *Geography* 86: 159–163.

Batterbury S and Ndi F. (2018) Land-grabbing in Africa. In: Binns T, Lynch K and Nel E (eds) *The Routledge Handbook of African Development*. London: Routledge, 573–582.

Bayliss K. (2003) Utility privatisation in Sub-Saharan Africa: A case study of water. *Journal of Modern African Studies* 507–531.

BBC. (2002) *Huge Turnout for Countryside March*. Available at: http://news.bbc.co.uk/1/hi/uk/2274129.stm.

BBC. (2008) *Wheat Prices Hit Record Highs*. Available at: www.bbc.co.uk/worldservice/learningenglish/newsenglish/witn/2008/02/080229_wheat.shtml.

BBC. (2011) *Bangor Farmer Tells Superstar to 'Cover Up'*. Available at: https://www.bbc.co.uk/news/uk-northern-ireland-15104707.

BBC. (2012) *Media Reaction to London 2012 Olympic Opening Ceremony*. Available at: www.bbc.co.uk/news/uk-19025686.

BBC. (2019) *Climate Change Food Calculator: What's Your Diet's Carbon Footprint?* Available at: www.bbc.co.uk/news/science-environment-46459714.

BBC. (2020) *Black Lives Matter: My Fight to Protest*. Available at: www.bbc.co.uk/iplayer/episode/p08jkwl2/newsbeat-documentaries-black-lives-matter-my-fight-to-protest.

BBC. (2020) *Black Lives Matter Protest Organiser Faces Abuse over Plan to Honour Minorities*. Available at: www.bbc.co.uk/news/newsbeat-55044173.

BBC. (2020) *Coronavirus: Second-Home Owners Urged to Stay Away*. Available at: https://www.bbc.co.uk/news/uk-wales-politics-51937753.

BBC. (2020) *Lydney Councillor Quits after Abuse for BLM Support*. Available at: www.bbc.co.uk/news/uk-england-gloucestershire-53473641.

Beauvoir S. (1972) *The Second Sex*. Harmondsworth: Penguin.

Bell D. (2006) Variations on the rural idyll. In: Cloke P, Goodwin M and Mooney P (eds) *Handbook of Rural Studies*. London: SAGE, 149–160.

Bell D and Jayne M. (2010) The creative countryside: Policy and practice in the UK rural cultural economy. *Journal of Rural Studies* 26: 209–218.

Bell D and Valentine G. (1995) Queer country: Rural lesbian and gay lives. *Journal of Rural Studies* 11: 113–122.

Bell M. (1994) *Childerley*. Chicago, IL: University of Chicago Press.

Belle E, Kingston N, Burgess N, Sandwith T, Ali N and MacKinnon K. (2018) *Protected Planet Report 2018*. Cambridge: United Nations Environment Programme.

Benson M. (2013) *The British in Rural France: Lifestyle Migration and the Ongoing Quest for a Better Way of Life*. Manchester: Manchester University Press.

Benson M and O'Reilly K. (2009) *Lifestyle Migration: Expectations, Aspirations and Experiences*. London: Routledge.

Bergman K. (2014) *Swedish Crime Fiction: The Making of Nordic Noir*. Italy: Mimesis.

Berry B. (1976) *Urbanization and Counter-Urbanization*. Beverly Hills, CA: SAGE.

Bertolini P. (2019) *Overview of Income and Non Income Rural Poverty in Developed Countries* Available at: www.un.org/development/desa/dspd/wp-content/uploads/sites/22/2019/03/bertolini-presentation-on-rural-poverty-developed-countries-2.pdf.

Bessiere J. (1998) Local development and heritage: Traditional food and cuisine as tourist attractions in rural areas. *Sociologia Ruralis* 38: 21.

Black Lives Matter. (2022) *Herstory*. Available at: https://blacklivesmatter.com/herstory/.

Blacksell M. (2005) A walk on the South West Coast Path: A view from the other side. *Transactions of the Institute of British Geographers* 30: 518–520.

Blackstock K, Innes A, Cox S, Smith A and Mason A. (2006) Living with dementia in rural and remote Scotland: Diverse experiences of people with dementia and their carers. *Journal of Rural Studies* 22: 161–176.

Block D and DuPuis E. (2001) Making the country work for the city: Von Thünen's ideas in geography, agricultural economics and the sociology of agriculture. *American Journal of Economics and Sociology* 60: 79–98.

Blunt A. (2006) Cultural geography: Cultural geographies of home. *Progress in Human Geography* 29: 505–515.

Blunt A and Dowling R. (2006) *Home*. London: Routledge.

Blunt A and Varley A. (2004) Geographies of home. *Cultural Geographies* 11: 3–6.

Boogaard B, Bock B, Oosting S, and Kroch E. (2010) Visiting a farm: An exploratory study of the social construction of animal farming in Norway and the Netherlands based on sensory perception. *The International Journal of Sociology of Agriculture and Food* 17: 24–50.

Bosworth G. (2010) Commercial counterurbanisation: An emerging force in rural economic development. *Environment and Planning A: Economy and Space* 42: 966–981.

Bosworth G and Bat Finke H. (2020) Commercial counterurbanisation: A driving force in rural economic development. *Environment and Planning A: Economy and Space* 52: 654–674.

Bouniol J. (2013) Scramble for land in Romania jeopardizes rural territories development. In: Franco J and Borras S Jr (eds) *Land Concentration, Land Grabbing and People's Struggles in Europe*. Amsterdam: Transnational Institute.

Bourdieu P. (1984) *Distinction: A Social Critique of the Judgement of Taste*. Cambridge, MA: Harvard University Press.

Boyle P and Halfacree K. (1998) *Migration into Rural Areas: Theories and Issues*. New York: John Wiley and Sons Ltd.

Brace C. (1999) Looking back: The Cotswolds and English national identity, c. 1890–1950. *Journal of Historical Geography* 25: 502–516.

Brace C. (2003) Rural mappings. *Country Visions* 49–72.

Bradford M and Kent A. (1993) *Understanding Human Geography: People and Their Changing Environments*. Oxford: Oxford University Press.

Brandth B. (1995) Rural masculinity in transition – gender images in tractor advertisements. *Journal of Rural Studies* 11: 123–133.

Brandth B. (2002) Gender identity in European family farming: A literature review. *Sociologia Ruralis* 42: 181–200.

Brandth B. (2006) Agricultural body-building: Incorporations of gender, body and work. *Journal of Rural Studies* 22: 17–27.

Brandth B and Haugen M. (2007) Gendered work in family farm tourism. *Journal of Comparative Family Studies* 38: 379–393plymouth.

Bressey C. (2009) Cultural archaeology and historical geographies of the black presence in rural England. *Journal of Rural Studies* 25: 386–395.

Brickell K. (2012) 'Mapping' and 'doing' critical geographies of home. *Progress in Human Geography* 36: 225–244.

Brickell K and Datta A. (2011) *Translocal Geographies*. Farnham: Ashgate Publishing.

Brittan G Jr. (2001) Wind, energy, landscape: Reconciling nature and technology. *Philosophy & Geography* 4: 169–184.

Brooks X. (2018) Road to nowhere: the new crop of writers unearthing the dark side of village life. *The Guardian* 3rd March 5–11.

Brown C, McMorran R and Price M. (2011) Rewilding – a new paradigm for nature conservation in Scotland? *Scottish Geographical Journal* 127: 288–314.

Brown D, Champion T, Coombes M, and Wymer C. (2015) The Migration-commuting nexus in rural England. A longitudinal analysis. *Journal of Rural Studies* 41: 118–128.

Brown D and Glasgow N. (2008) *Rural Retirement Migration*. New York: Springer Science & Business Media.

Browne K. (2009) Womyn's separatist spaces: Rethinking spaces of difference and exclusion. *Transactions of the Institute of British Geographers* 34: 541–556.

Browne K. (2011) Beyond rural idylls: Imperfect lesbian utopias at Michigan Womyn's music festival. *Journal of Rural Studies* 27: 13–23.

Brunori G and Bartolini F. (2019) The family farm. Model for the future or relic of the past? In: Shucksmith M and Brown D (eds) *Routledge International Handbook of Rural Studies*. London: Routledge.

Brunsdale M. (2016) *Encyclopedia of Nordic Crime Fiction: Works and Authors of Denmark, Finland, Iceland, Norway and Sweden Since 1967*. Jefferson, NC: McFarland.

Bryant L. (1999) The detraditionalization of occupational identities in farming in South Australia. *Sociologia Ruralis* 39: 236–261.

Bryant L and Garnham B. (2017) Bounded choices: The problematisation of longterm care for people ageing with an intellectual disability in rural communities. *Journal of Rural Studies* 51: 259–266.

Buchanan R. (1959) Some reflections on agricultural geography: Address to the Geographical Association. *Geography* 44: 1–13.

Buller H. (2004) Where the wild things are: The evolving iconography of rural fauna. *Journal of Rural Studies* 20: 131–141.

Buller H and Morris C. (2003) Farm animal welfare: A new repertoire of nature-society relations or modernism re-embedded? *Sociologia Ruralis* 43: 216–237.

Bunce M. (1994) *The Countryside Ideal: Anglo-American Images of Landscape*. London; New York: Routledge.

Burholt V and Dobbs C. (2012) Research on rural ageing: Where have we got to and where are we going in Europe? *Journal of Rural Studies* 28: 432–446.

Burke A and Jones A. (2019) The development of an index of rural deprivation: A case study of Norfolk, England. *Social Science & Medicine* 227: 93–103.

Burton R. (2004) Seeing through the 'good farmer's' eyes: Towards developing an understanding of the social symbolic value of 'productivist' behaviour. *Sociologia Ruralis* 44: 195.

Bushin N. (2009) Researching family migration decision-making: A children-in-families approach. *Population, Space and Place* 15: 429–443.

Butler J. (1993) *Bodies That Matter: On the Discursive Limits of Sex*. London: Routledge.

Butler R. (1999) The body. In: Cloke P, Crang M and Goodwin M (eds) *Introducing Human Geographies*. London: Arnold.

Butler R and Parr H. (1999) *Mind and Body Spaces: Geographies of Illness, Impairment, and Disability*. London; New York: Routledge.

Buttimer A and Mels T. (2006) *By Northern Lights: On the Making of Geography in Sweden*. Aldershot; Burlington, VT: Ashgate Publishing.

Byrne A, Edmondson R and Varley T. (2015) Arensberg and Kimball and anthropological research in Ireland. *Irish Journal of Sociology* 23: 22–61.

Byrne J and Wolch J. (2009) Nature, race, and parks: Past research and future directions for geographic research. *Progress in Human Geography* 33: 743–765.

Byrne R and Smith K. (2016) Modern slavery and agriculture. In: Donnermeyer JF (ed.) *The Routledge International Handbook of Rural Criminology*. London: Routledge.

Caines M. (2022) *If You Haven't Failed, You're Not Living Your Life*. Available at: https://www.telegraph.co.uk/connect/small-business/the-black-farmer-have-not-failed-not-living-life/.

Campbell H. (2000) The glass phallus: Pub(lic) masculinity and drinking in rural New Zealand. *Rural Sociology* 65: 562–581.

Campbell H and Bell M. (2000) The question of rural masculinities. *Rural Sociology* 65: 532–546.

Campbell H, Mayerfeld Bell M and Finney E. (2006) *Country Boys*. State College, PA: Pennsylvania State University.

Carmody P. (2013) *The Rise of the BRICS in Africa: The Geopolitics of South-South Relations*. London: Zed Books.

Carmody P and Ofori A. (2020) *Land Grabs*. Available at: https://planetgeogblog.wordpress.com/2019/12/16/land-grabs/.

Carrington, K., R. Hogg & M. Sozzo (2015) Southern Criminology. *The British Journal of Criminology*, 56, 1-20.

Carrington K, McIntosh A, Hogg R and Scott J. (2013) Rural masculinities and the internalisation of violence in agricultural communities. *International Journal of Rural Criminology* 2: 3–24.

Carson R. (1962) *Silent Spring*. Boston: Houghton Mifflin Harcourt.

Carter J and Hollinsworth D. (2009) Segregation and protectionism: Institutionalised views of Aboriginal rurality. *Journal of Rural Studies* 25: 414–424.

Castree N. (2012) Progressing physical geography. *Progress in Physical Geography: Earth and Environment* 36: 298–304.

Castree N. (2014) The Anthropocene and geography I: The back story. *Geography Compass* 8: 436–449.

Cater J and Jones T. (1989) *Social Geography*. London: Edward Arnold.

Ceccato V and Dolmen L. (2011) Crime in rural Sweden. *Applied Geography* 31: 119–135.

Ceccato V, Lundqvist P, Abraham J, Göransson E and Svennefelt C. (2022) Farmers, victimization, and animal rights activism in Sweden. *The Professional Geographer* 74: 350–363.

Cella M. (2017) Retrofitting rurality: Embodiment and emplacement in disability narratives. *Journal of Rural Studies* 51: 284–294.

Chakraborti N. (2010) Beyond 'passive Apartheid'? Developing policy and research agendas on rural racism in Britain. *Journal of Ethnic and Migration Studies* 36: 501–517.

Chakraborti N and Garland J. (2004) *Rural Racism*. Cullompton; Portland, OR: Willan Publisher.

Chakraborti N and Garland J. (2013) *Rural Racism*. London: Willan.

Champion A. (1989) *Counterurbanization: The Changing Pace and Nature of Population Deconcentration*. London: Arnold.

Champion T. (2001) Urbanization, suburbanization, counterurbanization and reurbanization. *Handbook of Urban Studies* 160: 1.

Champion T. (2009) Urban – rural differences in commuting in England: A challenge to the rural sustainability agenda? *Planning Practice & Research* 24: 161–183.

Chaney P and Sherwood K. (2000) The resale of right to buy dwellings: A case study of migration and social change in rural England. *Journal of Rural Studies* 16: 79–94.

Chapman C. (2004) *Silence at Ramscliffe: A Personal Account*. www.chrischapmanphotography.com.

Chapman C and Crowden J. (2005) *Silence at Ramscliffe: Foot and Mouth in Devon*. Oxford: Bardwell Press.

Charatsari C and Černič Istenič M. (2016) Gender, farming and rural social research: A relationship in flux. In: Shucksmith M and Brown D (eds) *Routledge International Handbook of Rural Studies*. London: Routledge, 470–481.

Cheeseman, J; Hays, D and Smith, A (2016) A Glance at the Age Structure and Labor Force Participation of Rural America https://www.census.gov/newsroom/blogs/random-samplings/2016/12/a_glance_at_the_age.html Accessed 21/02/2023

Cheshire L. (2016) Power and governance. In: Shucksmith M and Brown D (eds) *Routledge International Handbook of Rural Studies*. London: Routledge, 593–600.

Cheshire L, Higgins V and Lawrence G. (2006) *Rural Governance: International Perspectives*. London: Routledge.

Cheung Y, Spittal M, Pirkis J, Yip P (2012) Spatial analysis of suicide mortality in Australia: Investigation of metropolitan-rural-remote differentials of suicide risk across states/territories. *Social Science & Medicine* 75: 1460–1468.

Chipeniuk R. (2004) Planning for amenity migration in Canada. *Mountain Research and Development* 24: 327–335.

Chisholm M. (1962) *Rural Settlement and Land Use*. London: Hutchison.

Chiswell H and Wheeler R. (2016) 'As long as you're easy on the eye': Reflecting on issues of positionality and researcher safety during farmer interviews. *Area* 48: 229–235.

Christaller W. (1933) *Die zentralen Orte in Süddeutschland (The Central Places in Southern Germany)*. Jena: Gustav Fischer.

Chueh H and Lu Y. (2018) My dream life in a rural world: A nonfiction media representation of rural idyll in Taiwan. *Journal of Rural Studies* 59: 132–141.

Cloke H, Wetterhall F, He Y, Freer J and Pappenberger F. (2013) Modelling climate impact on floods with ensemble climate projections. *Quarterly Journal of the Royal Meteorological Society* 139: 282–297.

Cloke P. (1977) An index of rurality for England and Wales. *Regional Studies* 11: 31–46.

Cloke P. (1979) *Key Settlements in Rural Areas*. London: Methuen and Co.

Cloke P. (1980) New emphases for applied rural geography. *Progress in Human Geography* 4: 181–217.

Cloke P. (1983) *An Introduction to Rural Settlement Planning*. London: Methuen and Co.

Cloke P. (1989) Rural geography and political economy. *New Models in Geography* 1: 164–197.

Cloke P. (1993) On 'problems and solutions'. The reproduction of problems for rural communities in Britain during the 1980s. *Journal of Rural Studies* 9: 113–121.

Cloke P. (1994) (En)culturing political economy: A life in the day of a 'rural geographer'. In: Cloke P, Doel M, Matless D, Thrift N and Phillips M. (eds) *Writing the Rural: Five Cultural Geographies*. London: SAGE, 149–190.

Cloke P. (1996) Critical writing on rural studies: A short reply to Simon Miller. *Sociologia Ruralis* 36: 117–120.

Cloke P. (1997) Country backwater to virtual village? Rural studies and 'the cultural turn'. *Journal of Rural Studies* 13: 367–375.

Cloke P. (2003) *Country Visions*. Harlow; New York: Pearson/Prentice Hall.

Cloke P. (2004) *Practising Human Geography*. London; Thousand Oaks, CA: SAGE.

Cloke P. (2006a) Conceptualizing rurality. In: Cloke P, Marsden T and Mooney P (eds) *Handbook of Rural Studies*. London: SAGE, 18–28.

Cloke P. (2006b) Rurality and racialised others. In: Cloke P, Marsden T and Mooney P (eds) *Handbook of Rural Studies*. London: SAGE.

Cloke P. (2013) Rural landscapes. In: Johnson N, Schein R and Winders J (eds) *The Wiley-Blackwell Companion to Cultural Geography*. Malden: Wiley-Blackwell, 225–237.

Cloke P. (2014) Self-other. In: Cloke P, Crang P and Goodwin M (eds) *Introducing Human Geographies*. London: Routledge, 63–80.

Cloke P and Edwards G. (1986) Rurality in England and Wales 1981: A replication of the 1971 index. *Regional Studies* 20: 289–306.

Cloke P and Goodwin M. (1992) Conceptualizing countryside change: From post-Fordism to rural structured coherence. *Transactions of the Institute of British Geographers* 321–336.

Cloke P, Goodwin M, Milbourne P, and Thomas C. (1995a) Deprivation, poverty and marginalization in rural lifestyles in England and Wales. *Journal of Rural Studies* 11: 351–365.

Cloke P and Jones O. (2001) Dwelling, place, and landscape: An orchard in Somerset. *Environment and Planning A* 33: 649–666.

Cloke P and Little J. (1997) *Contested Countryside Cultures: Otherness, Marginalisation, and Rurality*. London; New York: Routledge.

Cloke P and Milbourne P. (1992) Deprivation and life-styles in rural Wales 2. Rurality and the cultural dimension. *Journal of Rural Studies* 8: 359–371.

Cloke P, Milbourne P and Thomas C. (1997) Living lives in different ways? Deprivation, marginalization and changing lifestyles in rural England. *Transactions of the Institute of British Geographers* 210–230.

Cloke P, Milbourne P and Widdowfield R. (2000) Homelessness and rurality: 'Out-of-place' in purified space? *Environment and Planning D: Society and Space* 18: 715–735.

Cloke P, Milbourne P and Widdowfield R. (2002) *Rural Homelessness: Issues, Experiences and Policy Responses*. Bristol: Policy Press.

Cloke P and Perkins H. (1998) "Cracking the canyon with the awesome foursome": Representations of adventure tourism in New Zealand. *Environment and Planning D: Society and Space* 16: 185–218.

Cloke P and Perkins H. (2005) Cetacean performance and tourism in Kaikoura, New Zealand. *Environment and Planning D: Society and Space* 23: 903–924.

Cloke P, Phillips M and Thrift N. (1995b) The new middle classes and the social construction of rural living. In: Butler T and Savage M (eds) *Social Change and the New Middle Classes*. London: UCL Press.

Cloke P and Thrift N. (1987) Intra-class conflict in rural areas. *Journal of Rural Studies* 3: 321–333.

Closs Stephens A. (2016) The affective atmospheres of nationalism. *Cultural Geographies* 23: 181–198.

Clout H. (1980) Rural settlements. *Progress in Human Geography* 4: 392–398.

Coldwell I. (2007) Young farmers, masculinities and the embodiment of farming practices in an Australian setting. *Rural Society* 17: 19–33.

Coley D, Howard M and Winter M. (2011) Food miles: Time for a re-think? *British Food Journal* 113: 919–934.

Collard R. (2012) Cougar – human entanglements and the biopolitical un/making of safe space. *Environment and Planning D: Society and Space* 30: 23–42.

Connell J and Gibson C. (2003) *Sound Tracks: Popular Music, Identity, and Place*. London; New York: Routledge.

Connell, R. (2014) Using southern theory: Decolonizing social thought in theory, research and application. *Planning Theory*, 13, 210-223.

Community Land Trust Network. (2022) *What Is a Community Land Trust (CLT)?* Available at: https://www.communitylandtrusts.org.uk/about-clts/what-is-a-community-land-trust-clt/.

Conradson D. (2003) Geographies of care: Spaces, practices, experiences. *Social & Cultural Geography* 4: 451–454.

Constantin C, Luminiţa C and Vasile A. (2017) Land grabbing: A review of extent and possible consequences in Romania. *Land Use Policy* 62: 143–150.

Convery I, Bailey C, Mort M, and Baxter J. (2005) Death in the wrong place? Emotional geographies of the UK 2001 foot and mouth disease epidemic. *Journal of Rural Studies* 21: 99–109.

Coombes M, Dalla Longga R and Raybould S. (1989) Counterurbanisation in Britain and Italy: A comparative critique of the concept, causation and evidence. *Progress in Planning* 32: 1–70.

Cooper D and Gregory I. (2011) Mapping the English Lake District: A literary GIS. *Transactions of the Institute of British Geographers* 36: 89–108.

Coppock J. (1971) *An Agricultural Geography of Great Britain*. London: G Bell and Sons.

Cosgrove D. (1984) *Social Formation and Symbolic Landscape*. Madison, WI: University of Wisconsin Press.

Cosgrove D. (1985) Prospect, perspective and the evolution of the landscape idea. *Transactions of the Institute of British Geographers* 10: 45–62.

Cosgrove D. (1995) Wild ideas. In: Rothenburg D (ed.) *Wild Ideas*. Minneapolis, MN: University of Minnesota, 27–41.

Cosgrove D. (1998) *Social Formation and Symbolic Landscape*. Madison, WI: University of Wisconsin Press.

Cotula L. (2012) The international political economy of the global land rush: A critical appraisal of trends, scale, geography and drivers. *The Journal of Peasant Studies* 39: 649–680.

Couper P. (2014) *A Student's Introduction to Geographical Thought: Theories, Philosophies, Methodologies*. London; Thousand Oaks, CA: SAGE.

Couper P and Yarwood R. (2012) Confluences of human and physical geography research on the outdoors: An introduction to the special section on 'Exploring the outdoors'. *Area* 44: 2–6.

Cox C, Moore P and Ladle R. (2016) *Biogeography: An Ecological and Evolutionary Approach* (9th ed.). New York: Wiley.

Coy M, Kelly L and Foord J. (2009) *Map of Gaps: the Postcode Lottery of Violence against Women Support Services in Britain*. London: End Violence against Women.

Crang M. (1999) Nation, region and homeland: History and tradition in Dalarna, Sweden. *Ecumene* 6: 447–470.

Crang M. (2015) Representations-reality. In: Cloke P, Goodwin M and Crang P (eds) *Introducing Human Geographies* (3rd ed.). London: Routledge, 130–144.

Crang M and Tolia-Kelly D. (2010) Nation, race, and affect: Senses and sensibilities at National Heritage sites. *Environment and Planning A: Economy and Space* 42: 2315–2331.

Crenshaw K. (1990) Mapping the margins: Intersectionality, identity politics, and violence against women of color. *Stanford Law Review* 43: 1241.

Cresswell T. (2009) The prosthetic citizen: New geographies of citizenship. *Political Power and Social Theory* 20: 259–273.

Cresswell T. (2010) Towards a politics of mobility. *Environment and Planning D: Society and Space* 28: 17–31.

Crook M. (2020) Forward. In: Innes K and Norcup J (eds) *Landscapes of Detectorists*. Axminister: Uniform Books, 8–12.

Crutzen P. (2006) The "Anthropocene". In: Ehlers E and Krafft T (eds) *Earth System Science in the Anthropocene*. New York: Springer, 13–18.

Cummins L, First R and Toomey B. (1998) Comparisons of rural and urban homeless women. *Affilia* 13: 435–453.

Cunneen C. (2001) Conflict, politics and crime: Aboriginal communities and the police. In: *Conflict, Politics and Crime: Aboriginal Communities and the Police*. Crows Nest: Allen and Unwin.

Curry G, Koczberski G and Selwood J. (2001) Cashing out, cashing in: Rural change on the south coast of Western Australia. *Australian Geographer* 32: 109–124.

Curry N and Owen S. (2009) Rural planning in England: A critique of current policy. *The Town Planning Review* 575–595.

Dacey M. (1965) The geometry of central place theory. *Geografiska Annaler: Series B, Human Geography* 47: 111–124.

Dafydd Jones R. (2010) Islam and the rural landscape: Discourses of absence in west Wales. *Social & Cultural Geography* 11: 751–768.

Daniels S. (1993) *Fields of Vision: Landscape Imagery and National Identity in England and the United States*. Cambridge: Polity Press.

Daniels S and Cosgrove D. (1988) *The Iconography of Landscape*. Cambridge: Cambridge University Press.

Dartmoor National Park Authority. (2019) *Settlement Profile: Princetown*. Parke: Dartmoor National Park Authority.

Davidoff L. (1995) *Worlds Between: Historical Perspectives on Gender and Class*. Hove: Psychology Press.

Davies R and Partington R. (2018) More than 25% of UK pubs have closed since 2001. *The Guardian*. Available at: https://www.theguardian.com/business/2018/nov/26/uk-pub-closures-financial-crisis-birmingham-ons-figures.

Dawney L. (2007) *Supporting Integration of Migrants and Seasonal Workers in the Diocese of Hereford. Report to the Diocese of Hereford*. Hereford: Council for Social Responsibility.

Dax T, Strahl W, Kirwan J, and Maye D. (2016) The Leader programme 2007–2013: Enabling or disabling social innovation and neo-endogenous development? Insights from Austria and Ireland. *European Urban and Regional Studies* 23: 56–68.

Day G. (1998) A community of communities? Similarity and difference in Welsh rural community studies. *The Economic and Social Review* 29: 223–257.

Deary H and Warren C. (2017) Divergent visions of wildness and naturalness in a storied landscape: Practices and discourses of rewilding in Scotland's wild places. *Journal of Rural Studies* 54: 211–222.

Delanty G. (2018) *Community*. London: Routledge.

de la O Campos A, Villani C, Davis B, and Takag M. (2018) *Ending Extreme Poverty in Rural Areas Sustaining Livelihoods to Leave No One Behind*. Rome: Food and Agricultural Org.

del Casino Jr V. (2009) *Social Geography: A Critical Introduction*. London: John Wiley & Sons.

Department for Environment, Food and Rural Affairs. (2018) *Organic Farming Statistics*. London: DEFRA.

Department for Environment, Food and Rural Affairs. (2019) *Statistical Digest for Rural England*. London: DEFRA.

Department for Environment, Food and Rural Affairs. (2021) *Statistical Digest for Rural England 2020*. London: DEFRA.

Department for Environment, Food and Rural Affairs. (2022) *Statistical Digest for Rural England*. London: DEFRA.

Derickson K. (2017) Urban geography II: Urban geography in the age of Ferguson. *Progress in Human Geography* 41: 230–244.

Desforges L, Jones R and Woods M. (2005) New geographies of citizenship. *Citizenship Studies* 9: 439–451.

Devine-Wright P. (2005) Beyond NIMBYism: Towards an integrated framework for understanding public perceptions of wind energy. *Wind Energy* 8: 125–139.

Devine-Wright P, Smith J and Batel S. (2019) "Positive parochialism", local belonging and ecological concerns: Revisiting Common Ground's Parish Maps project. *Transactions of the Institute of British Geographers* 44: 407–421.

Dibden J, Gibbs D and Cocklin C. (2013) Framing GM crops as a food security solution. *Journal of Rural Studies* 29: 59–70.

Dilley R and Scraton S. (2010) Women, climbing and serious leisure. *Leisure Studies* 29: 125–141.

Dinkovski N. (2022) *The Black Farmer: Brand of Plenty*. Available at: https://www.foodmanufacture.co.uk/Article/2016/08/10/The-Black-Farmer#.

Dimou, E. (2021) Decolonizing Southern Criminology: What Can the "Decolonial Option" Tell Us About Challenging the Modern/Colonial Foundations of Criminology? *Critical Criminology*, 1-20.

Dixon J and Banwell C. (2019) Supermarketisation and rural society futures. In: Shucksmith M and Brown D (eds) *Routledge International Handbook of Rural Studies*. London: Routledge.

Dodgshon R. (1987) *European Past*. Basingstoke: Macmillan International Higher Education.

Dodgshon R and Olsson G. (2006) Heather moorland in the Scottish Highlands: The history of a cultural landscape, 1600–1880. *Journal of Historical Geography* 32: 21–37.

Donaldson C, Gregory I and Taylor J. (2017) Locating the beautiful, picturesque, sublime and majestic: Spatially analysing the application of aesthetic terminology in descriptions of the English Lake District. *Journal of Historical Geography* 56: 43–60.

Donnermeyer J and Barclay E. (2005) The policing of farm crime. *Police Practice and Research* 6: 3–17.

Dorling D and Gunnell D. (2003) Suicide: The spatial and social components of despair in Britain 1980–2000. *Transactions of the Institute of British Geographers* 28: 442–460.

Drysdale R. (1991) Aged migration to coastal and inland centres in NSW. *Australian Geographical Studies* 29: 268–284.

Dufhues T, Möllers J, Traikova D, Buchenrieder G and Runschke D. (2020) Why villagers stay put – A structural equation model on staying intentions. *Journal of Rural Studies* 81: 345–357.

Earle F. (1994) *A Time to Travel? An Introduction to Britain's Newer Travellers*. Eyemouth: Enabler Publications.

Edensor T. (2000) Walking in the British Countryside: Reflexivity, embodied practices and ways to escape. *Body & Society* 6: 81–106.

Edensor T. (2002) *National Identity, Popular Culture and Everyday Life*. London: Routledge.

Edensor T. (2006) Performing rurality. In: Cloke P, Marsden T and Mooney P (eds) *Handbook of Rural Studies*. London: SAGE, 484–495.

Edwards B, Goodwin M, Pemberton S, and Woods M. (2001) Partnerships, power, and scale in rural governance. *Environment and Planning C: Government and Policy* 19: 289–310.

Emmanuel-Jones, W (2023) About me https://www.theblackfarmer.com/about/#about-me Accessed 20/02/2023

Eimermann M. (2013) Lifestyle migration to the North: Dutch families and the decision to move to rural Sweden. *Population, Space and Place* 21: 68–85.

Eltham D, Harrison G and Allen S. (2008) Change in public attitudes towards a Cornish wind farm: Implications for planning. *Energy Policy* 36: 23–33.

Enticott G. (2001) Calculating nature: The case of badgers, bovine tuberculosis and cattle. *Journal of Rural Studies* 17: 149–164.

Enticott G. (2008) The spaces of biosecurity: Prescribing and negotiating solutions to bovine tuberculosis. *Environment and Planning A: Economy and Space* 40: 1568–1582.

Essex S, Gilg A, Yarwood R, Smithers J and Wilson R. (2005) *Rural Change and Sustainability: Agriculture, the Environment and Communities*. Wallingford: CABI.

European Commission. (2006) *The Leader Approach*. Luxembourg: European Communities.

European Commission. (2022) *Rural Development*. Available at: https://agriculture.ec.europa.eu/common-agricultural-policy/rural-development_en.

European Commission (2023) *Is my region covered?* https://ec.europa.eu/regional_policy/policy/how/is-my-region-covered_en

European Parliament. (2020) *LEADER Experiences – Lessons Learned and Effectiveness of EU Funds for Rural Development*. Available at: http://elard.eu/wp-content/uploads/2020/04/20181113_Briefing-formatted-Leader-Experience_AP.pdf.

European Rural Network Assembly. (2015) *LEADER 2007–2013 Implementation Update*. Available at: https://enrd.ec.europa.eu/sites/enrd/files/uploaded-files/leader1_kjasinska.pdf.

European Union. (2015) *Extent of Farmland Grabbing in the EU*. Brussels: European Union.

Eurosat. (2020a) *Housing Statistics*. Available at: https://ec.europa.eu/eurostat/statistics-explained/index.php/Housing_statistics#Tenure_status.

Eurosat. (2020b) *Urban-rural Typology*. Available at: https://ec.europa.eu/eurostat/web/rural-development/methodology.

Evans N and Yarwood R. (1995) Livestock and landscape. *Landscape Research* 20: 141–146.

Evans N and Yarwood R. (1998) New places for "old spots": The changing geographies of domestic livestock animals. *Society & Animals* 6: 137–165.

Evans N. (2013) Strawberry fields forever? Conflict over neo-productivist Spanish polytunnel technology in British agriculture. *Land Use Policy* 35: 61–72.

Evans N, Morris C and Winter M. (2002) Conceptualizing agriculture: A critique of post-productivism as the new orthodoxy. *Progress in Human Geography* 26: 313–332.

Everingham J. (2016) Transformations of rural society and environments by extraction of mineral and energy resources. In: Shucksmith M and Brown D (eds) *Routledge International Handbook of Rural Studies*. London: Routledge, 272–298.

Falk I and Kilpatrick S. (2000) What is social capital? A study of interaction in a rural community. *Sociologia Ruralis* 40: 87–110.

FAO. (2021) *FAOSAT*. Available at: https://www.fao.org/faostat/en/.

Ferbrache F and Yarwood R. (2015) Britons abroad or European citizens? The negotiation of (trans)national space and citizenship by British migrants in France. *Geoforum* 62: 73–83.

Fernando P. (1998) Gender and rural transport. *Gender, Technology and Development* 2: 63–80.

Feyen L, Dankers R, Bódis K, Salamon P and Barredo J. (2012) Fluvial flood risk in Europe in present and future climates. *Climatic Change* 112: 47–62.

Fielding A. (1989) Migration and urbanization in Western Europe since 1950. *The Geographical Journal* 155: 60–69.

Fish R. (2007) *Cinematic Countrysides*. Manchester: Manchester University Press.

Fitchen J. (1992) On the edge of homelessness: Rural poverty and housing insecurity 1. *Rural Sociology* 57: 173–193.

Fletschner D and Kenney L. (2014) Rural women's access to financial services: Credit, savings, and insurance. In: Quisumbing A, Meinzen-Dick R, Raney T, Croppenstedt A, Behrman J and Peterman A. (eds) *Gender in Agriculture: Closing the Knowledge Gap*. Dordrecht: Springer, 187–208.

Food and Agriculture Organization of the United Nations (FAO). (2009) *How to Feed the World in 2050*. Rome: FAO.

Food and Agriculture Organization of the United Nations (FAO). (2014) *Agriculture Organization of the United Nations (2014) State of the World's Forests: Enhancing the Socioeconomic Benefits from Forests*. Rome: FAO.

Forrest R and Murie A. (1992) Change on a rural council estate: An analysis of dwelling histories. *Journal of Rural Studies* 8: 53–65.

Forsyth N. (2012) The naked rambler: The man prepared to go to prison for nudity. *The Guardian*. Available at: www.theguardian.com/lifeandstyle/2012/mar/23/naked-rambler-prison.

Frankenburg R. (1966) *Communities in Britain*. London: Pelican.

Friedmann H and McMichael P. (1989) The rise and decline of national agricultures, 1870 to the present. *Sociologia Ruralis* 29: 93–117.

Fyfe N and Milligan C. (2003) Out of the shadows: Exploring contemporary geographies of voluntarism. *Progress in Human Geography* 27: 397–413.

Gallaher C. (2016) Placing the militia occupation of the Malheur National Wildlife Refuge in Harney County, Oregon. *ACME: An International E-Journal for Critical Geographies* 15.

Gallent N. (2007) Second homes, community and a hierarchy of dwelling. *Area* 39: 97–106.

Gallent N. (2013) Re-connecting 'people and planning': Parish plans and the English localism agenda. *The Town Planning Review* 371–396.

Gallent N. (2014) The social value of second homes in rural communities. *Housing, Theory and Society* 31: 174–191.

Gallent N. (2019) Situating rural areas in contemporary housing access debates in England – A comment. *Planning Practice & Research* 34: 489–497.

Gallent N. (2020) COVID-19 and the flight to second homes. *Town & Country Planning* 89: 141–144.

Gallent N, Hamiduddin I, Juntti M, and Shaw D. (2015) *Introduction to Rural Planning: Economies, Communities and Landscapes*. New York: Routledge.

Gallent N, Hamiduddin I, Juntti M, and Stirling P. (2018) *New Money in Rural Areas: Land Investment in Europe and Its Place Impacts*. New York: Springer.

Gallent N, Hamiduddin I, Kelsey J, et al. (2020) Housing access and affordability in rural England: Tackling inequalities through upstream reform or downstream intervention? *Planning Theory & Practice* 21: 531–551.

Gallent N, Juntti M, Kidd S, and Shaw D. (2008a) *Introduction to Rural Planning*. London: Routledge.

Gallent N, Mace A and Tewdwr-Jones M. (2003a) Dispelling a myth? Second homes in rural Wales. *Area* 35: 271–284.

Gallent N, Morphet J and Tewdwr-Jones M. (2008b) Parish plans and the spatial planning approach in England. *Town Planning Review* 79: 1–30.

Gallent N and Scott M. (2017) *Rural Planning and Development: Critical Concepts in Built Environment*. Abingdon; Oxon; New York: Routledge.

Gallent N, Shucksmith M and Tewdwr-Jones M. (2003b) *Housing in the European Countryside: Rural Pressure and Policy in Western Europe*. London; New York: Routledge.

Gans H. (1962) *Urban Villagers*. New York: The Free Press.

García E and Valverde F. (2009) La inversión en los programas de desarrollo rural. Su reparto territorial en la provincia de Granada/Investment in rural development programmes. Its territorial distribution in the province of Granada. *Anales de Geografía de la Universidad Complutense* 37.

Garrod B, Wornell R and Youell R. (2006) Re-conceptualising rural resources as countryside capital: The case of rural tourism. *Journal of Rural Studies* 22: 117–128.

Gasson R, Crow G, Errington A, Hutson J, Marsden T and Winter M (1988) The farm as a family business: A review. *Journal of Agricultural Economics* 39: 1–41.

George W. (1962) *Animal Geography*. London: Heinemann.

Gere C. (2019) *I Hate the Lake District*. London: Goldsmiths Press.

Gertel J and Sippel R. (2019) The financialisation of agriculture and food. In: Shucksmith M and Brown D (eds) *Routledge International Handbook of Rural Studies*. London: Routledge, 215–226.

Gibb R. (2004) Developing countries and market access: The bitter-sweet taste of the European Union's sugar policy in southern Africa. *The Journal of Modern African Studies* 42: 563–588.

Gibson C and Davidson D. (2004) Tamworth, Australia's 'country music capital': Place marketing, rurality, and resident reactions. *Journal of Rural Studies* 20: 387–404.

Gibson-Graham J. (1996) *The End of Capitalism (as We Knew It): A Feminist Critique of Political Economy*. Cambridge; Oxford: Blackwell.

Giddings R and Yarwood R. (2005) Growing up, going out and growing out of the countryside: Childhood experiences in rural England. *Children's Geographies* 3: 101–114.

Gieling J, Vermeij L and Haartsen T. (2017) Beyond the local-newcomer divide: Village attachment in the era of mobilities. *Journal of Rural Studies* 55: 237–247.

Gilbert M, Nicolas G, Cinardi G, Van Boeckel T, Vanwambeke S, Wint W and Robinson T. (2018) Global distribution data for cattle, buffaloes, horses, sheep, goats, pigs, chickens and ducks in 2010. *Scientific Data* 5: 180227.

Gilg A. (1991) Book review: The child in the country, Colin Ward. *Area* 23: 90.

Gill, A (2005) Admiring the Scenery *The Sunday Times* 29th May 2005 https://www.thetimes.co.uk/article/admiring-the-scenery-jd5j26655xj Accessed 13/02/2023

Gillespie K and Collard R. (2015) *Critical Animal Geographies: Politics, Intersections, and Hierarchies in a Multispecies World*. London; New York: Routledge, Taylor & Francis Group.

Gillon C. (2014) Amenity migrants, animals and ambivalent natures: More-than-human encounters at home in the rural residential estate. *Journal of Rural Studies* 36: 262–272.

Gkartzios M. (2013) 'Leaving Athens': Narratives of counterurbanisation in times of crisis. *Journal of Rural Studies* 32: 158–167.

Gkartzois M and Lowe P. (2019) Revisiting neo-endogenous rural development. In: Scott M, Gallent N and Gkartzois M (eds) *The Routledge Companion to Rural Planning*. London: Routledge, 159–169.

Gkartzios M and Shucksmith M. (2015) 'Spatial anarchy' versus 'spatial apartheid': Rural housing ironies in Ireland and England. *Town Planning Review* 86: 53–73.

Glasgow N and Brown D. (2012) Rural ageing in the United States: Trends and contexts. *Journal of Rural Studies* 28: 422–431.

Glover J. (2019) *Landscapes Review*. London: DEFRA.

Gold J and Revill G. (2004) *Representing the Environment*. London; New York: Routledge.

Goodman M, Maye D and Holloway L. (2010) *Ethical Foodscapes? Premises, Promises, and Possibilities*. London: SAGE.

Goodwin M. (1998) The governance of rural areas: Some emerging research issues and agendas. *Journal of Rural Studies* 14: 5–12.

Gorman-Murray A, Waitt G and Gibson C. (2012) Chilling out in 'cosmopolitan country': Urban/rural hybridity and the construction of Daylesford as a 'lesbian and gay rural idyll'. *Journal of Rural Studies* 28: 69–79.

Gray D, Shaw J and Farrington J. (2006) Community transport, social capital and social exclusion in rural areas. *Area* 38: 89–98.

Gray J. (1996) Cultivating farm life on the borders: Scottish Hill sheep farms and the European Community. *Sociologia Ruralis* 36: 27–50.

Gray J. (2014) Hefting onto place: Intersecting lives of humans and sheep on Scottish Hills landscape. *Anthrozoös* 27: 219–234.

Greater London National Park City. (2015) *National Park City: A Proposal to Make Greater London the World's First National Park City*. London: Greater London National Park City.

Green A. (2016) Changing dynamics of rural labour markets. In: Shucksmith M and Brown D (eds) *Routledge International Handbook of Rural Studies*. London: Routledge, 143–153.

Grigg D. (1984) *An Introduction to Agricultural Geography*. London: Unwin Hyman.

Grove K. (2013) Hidden transcripts of resilience: Power and politics in Jamaican disaster management. *Resilience* 1: 193–209.

Grove K. (2018) *Resilience*. New York: Routledge.

Gullino P and Larcher F. (2013) Integrity in UNESCO World Heritage Sites. A comparative study for rural landscapes. *Journal of Cultural Heritage* 14: 389–395.

Hajjar R, Oldekop J, Cronkleton P, Newton P, Russell A and Zhou W. (2021) A global analysis of the social and environmental outcomes of community forests. *Nature Sustainability* 4: 216–224.

Halfacree K. (2001) Constructing the object: Taxonomic practices, 'counterurbanisation' and positioning marginal rural settlement. *International Journal of Population Geography* 7: 395–411.

Halfacree K. (2006) From dropping out to leading on? British counter-cultural back-to-the-land in a changing rurality. *Progress in Human Geography* 30: 309–336.

Halfacree K. (2007) Trial by space for a 'radical rural': Introducing alternative localities, representations and lives. *Journal of Rural Studies* 23: 125–141.

Halfacree K. (2008) To revitalise counterurbanisation research? Recognising an international and fuller picture. *Population, Space and Place* 14: 479–495.

Halfacree K. (2009) 'Glow worms show the path we have to tread': The counterurbanisation of Vashti Bunyan. *Social & Cultural Geography* 10: 771–789.

Halfacree K. (2012) Heterolocal identities? Counter-urbanisation, second homes, and rural consumption in the era of mobilities. *Population, Space and Place* 18: 209–224.

Halfacree K. (2018) Hope and repair within the Western Skyline? Americana Music's rural heterotopia. *Journal of Rural Studies* 63: 1–14.

Halfacree K. (1993) Locality and social representation: Space, discourse and alternative definitions of the rural. *Journal of Rural Studies* 9: 23–37.

Halfacree K. (1994) The importance of the rural in the constitution of counterurbanization – evidence from England in the 1980s. *Sociologia Ruralis* 34: 164–189.

Halfacree K. (1995) Talking about rurality: Social representations of the rural as expressed by residents of six English parishes. *Journal of Rural Studies* 11: 1–20.

Halfacree K. (1996) Out of place in the country: Travellers and the "rural idyll". *Antipode* 28: 42.

Halfacree K and Rivera M. (2012) Moving to the countryside . . . and staying: Lives beyond representations. *Sociologia Ruralis* 52: 92–114.

Hall T. (2020) After the rural idyll: Representations of the British countryside as a non-idyllic environment. *Geography* 105: 6–17.

Hall T and Barrett H. (2018) *Urban Geography*. London: Routledge.

Hall T and Hubbard P. (1996) The entrepreneurial city: New urban politics, new urban geographies? *Progress in Human Geography* 20: 153–174.

Halliday J and Little J. (2001) Amongst women: Exploring the reality of rural childcare. *Sociologia Ruralis* 41: 423–437.

Halseth G. (2018) The changing nature of resource economies: A focus on the example of forestry. In: Shucksmith M and Brown D (eds) *Routledge International Handbook of Rural Studies*. London: Routledge, 108–119.

Haraway D. (2013) *Simians, Cyborgs, and Women: The Reinvention of Nature*. London: Routledge.

Harper S. (1989) The British rural community: An overview of perspectives. *Journal of Rural Studies* 5: 161–184.

Harper S. (2005) Contesting later life. In: Cloke P and Little J. (eds) *Contested Countryside Cultures*. London: Routledge.

Harries A. (2020) '"When I get up it just goes to shit" – unearthing the everyday vertical landscapes of the Dectorists'. In: Innes K and Norcup J (eds) *Landscapes of Detectorists*. Axminster: Uniform Books, 58–77.

Harrington L. (2018) Alternative and virtual rurality: Agriculture and the countryside as embodied in American imagination. *Geographical Review* 108: 250–273.

Harrington V and O'Donoghue D. (1998) Rurality in England and Wales 1991: A replication and extension of the 1981 rurality index. *Sociologia Ruralis* 38: 178–203.

Hart J. (1982) The highest form of the geographer's art. *Annals of the Association of American Geographers* 72: 1–29.

Hart J. (1998) *The Rural Landscape*. Baltimore, MD: John Hopkins University Press.

Hart J. (2003) *The Changing Scale of American Agriculture*. Charlottesville: University of Virginia Press.

Harvey D. (1973) *Social Justice and the City*. London: Edward Arnold.

Harvey D. (1989) *The Urban Experience*. Baltimore, MD: John Hopkins University Press.

Harvey D. (1990) Book review: New models in geography: The political economy perspective. *Transactions of the Institute of British Geographers* 15: 376–377.

Harvey D. (1963) Locational change in the Kentish Hop Industry and the analysis of land use patterns. *Transactions and Papers (Institute of British Geographers)* 123–144.

Haugen M and Villa M. (2006) Rural Idylls or boring places? In: Bock B and Shortall S (eds) *Rural Gender Relations: Issues and Case Studies*. Wallingford: CABI, 181–196.

Healey M and Ilbery B. (1985) *The Industrialization of the Countryside*. Norwich: Geo Books.

Heinonen J and Junnila S. (2011) A carbon consumption comparison of rural and urban lifestyles. *Sustainability* 3: 1234–1249.

Heley J. (2010) The new squirearchy and emergent cultures of the new middle classes in rural areas. *Journal of Rural Studies* 26: 321–331.

Heming L, Waley P and Rees P. (2001) Reservoir resettlement in China: Past experience and the Three Gorges Dam. *Geographical Journal* 167: 195–212.

Henderson S and Hoggart K. (2003) Ruralities and gender divisions of labour in Eastern England. *Sociologia Ruralis* 43: 349–378.

Henry M, Mahathey A, Morrill T, Robinson A, Shivji A and Watt R. (2018) *The 2018 Annual Homeless Assessment Report (AHAR) to Congress*. Washington, DC: The U.S. Department of Housing and Urban Development.

Herington J. (1984) *The Outer City*. New York: HarperCollins.

Her Majesty's Inspectorate of Constabularies. (2015) *Crime and Policing Comparator*. Available at: www.justiceinspectorates.gov.uk/mhim/crime-and-policing-comparator.

Herman A. (2015) Enchanting resilience: Relations of care and people–place connections in agriculture. *Journal of Rural Studies* 42: 102–111.

Herman A, Lahdesmaki M and Siltaoja M. (2018) Placing resilience in context: Investigating the changing experiences of Finnish organic farmers. *Journal of Rural Studies* 58: 112–122.

Herod A. (2011) *Scale*. London; New York: Routledge.

Hesse R, Allee W and Schmidt K. (1937) *Ecological Animal Geography*. London: John Wiley & Sons.

Hetherington K. (2000) *New Age Travellers: Van Loads of Uproarious Humanity*. London: Cassell.

Higgins V, Bryant M, Howell A, and Battersby J. (2017) Ordering adoption: Materiality, knowledge and farmer engagement with precision agriculture technologies. *Journal of Rural Studies* 55: 193–202.

Higgs G. (2003) *Rural Services and Social Exclusion*. London: Pion.

Higgs G and White S. (1997) Changes in service provision in rural areas. Part 1: The use of GIS in analysing accessibility to services in rural deprivation research. *Journal of Rural Studies* 13: 441–450.

Hill R. (2020) Aptitude or adaptation: What lies at the root of terroir? *The Geographical Journal* 186: 346–350.

Hinchliffe S. (2001) Indeterminacy in-decisions – science, policy and politics in the BSE (Bovine Spongiform Encephalopathy) crisis. *Transactions of the Institute of British Geographers* 26: 182–204.

Hinchliffe S, Allen J, Lavau S, Bingham, N and Carter S (2013) Biosecurity and the topologies of infected life: From borderlines to borderlands. *Transactions of the Institute of British Geographers* 38: 531–543.

Hockey J, Penhale B and Sibley D. (2005) Environments of memory: Home space, later life and grief. In: Davidson J, Bondi L and Smith M (eds) *Emotional Geographies*. Aldershot: Ashgate, 135–146.

Hodge I. (1996) On penguins on icebergs: The rural white paper and the assumptions of rural policy. *Journal of Rural Studies* 12: 331–337.

Hoey B. (2005) From pi to pie: Moral narratives of noneconomic migration and starting over in the postindustrial Midwest. *Journal of Contemporary Ethnography* 34: 586–624.

H. O. F. (1911) Zoogeography. *The Geographical Journal* 38: 413–419.

Hogg R and Carrington K. (2003) Violence, spatiality and other rurals. *Australian & New Zealand Journal of Criminology* 36: 293–319.

Hoggart K. (1988) Not a definition of rural. *Area* 20: 35–40.

Hoggart K. (1990) Let's do away with rural. *Journal of Rural Studies* 6: 245–257.

Hoggart K. (1998) Rural cannot equal middle class because class does not exist? *Journal of Rural Studies* 14: 381–386.

Hoggart K and Buller H. (1995) British home owners and housing change in rural France. *Housing Studies* 10: 179–198.

Hoggt R and Carrington K. (1998) Crime, rurality and community. *Australian & New Zealand Journal of Criminology* 31: 160–181.

Holloway L. (1999) Understanding climate change and farming: Scientific and farmers' constructions of 'global warming' in relation to agriculture. *Environment and Planning A: Economy and Space* 31: 2017–2032.

Holloway L. (2001) Pets and protein: Placing domestic livestock on hobby-farms in England and Wales. *Journal of Rural Studies* 17: 293–307.

Holloway L. (2002) Smallholding, hobby-farming, and commercial farming: Ethical identities and the production of farming spaces. *Environment and Planning A: Economy and Space* 34: 2055–2070.

Holloway L. (2004) Showing and telling farming: Agricultural shows and re-imaging British agriculture. *Journal of Rural Studies* 20: 319–330.

Holloway L, Bear C and Wilkinson K. (2014) Re-capturing bovine life: Robot–cow relationships, freedom and control in dairy farming. *Journal of Rural Studies* 33: 131–140.

Holloway L, Bear C and Wilkinson K. (2014) Re-capturing bovine life: Robot – cow relationships, freedom and control in dairy farming. *Journal of Rural Studies* 33: 131–140.

Holloway L, Morris C, Gilna B, and Gibbs D. (2009) Biopower, genetics and livestock breeding: (Re)constituting animal populations and heterogeneous biosocial collectivities. *Transactions of the Institute of British Geographers* 34: 394–407.

Holloway S. (2004) Rural roots, rural routes: Discourses of rural self and travelling other in debates about the future of Appleby New Fair, 1945–1969. *Journal of Rural Studies* 20: 143–156.

Holloway S. (2003) Outsiders in rural society? Constructions of rurality and nature – society relations in the racialisation of English Gypsy-Travellers, 1869–1934. *Environment and Planning D: Society and Space* 21: 695–715.

Holloway S. (2007) Burning issues: Whiteness, rurality and the politics of difference. *Geoforum* 38: 7–20.

Hoogendoorn G and Visser G. (2015) Focusing on the 'blessing' and not the 'curse' of second homes: Notes from South Africa. *Area* 47: 179–184.

Hopkins P. (2018) Feminist geographies and intersectionality. *Gender, Place & Culture* 25: 585–590.

Hopkins P and Noble G. (2009) Masculinities in place: Situated identities, relations and intersectionality. *Social & Cultural Geography* 10: 811–819.

Hopkins R. (2008) *The Transition Handbook: From Oil Dependency to Local Resilience*. Cambridge: Green Books.

Horton J. (2008) Producing Postman Pat: The popular cultural construction of idyllic rurality. *Journal of Rural Studies* 24: 389–398.

Hoskins W. (1955) *The Making of the English Landscape*. London: Penguin Books.

Howarth R, Ingraffea A and Engelder T. (2011) Should fracking stop? *Nature* 477: 271–275.

Hubbard P. (2005) 'Inappropriate and incongruous': Opposition to asylum centres in the English countryside. *Journal of Rural Studies* 21: 3–17.

Hughes A. (1997) Women and rurality: Gendered experiences of 'community' in village life. In: Milbourne P (ed.) *Revealing Rural Others: Representation, Power and Identity in the British Countryside*. London: Pinter.

Hughes E. (1993) *Jigso: Five Years of Community Action, Past Achievement and Future Developments*. Aberystwyth: Rural Survey Research Unit.

Hunsberger C, Corbera E, Borras Jr S, et al. (2017) Climate change mitigation, land grabbing and conflict: Towards a landscape-based and collaborative action research agenda. *Canadian Journal of Development Studies/Revue canadienne d'études du développement* 38: 305–324.

Hunter L, Boardman J and Onge J. (2005) The association between natural amenities, rural population growth, and long-term residents' economic well-being. *Rural Sociology* 70: 452–469.

Hunter L, Talbot C, Connor D, Connor D, Counterman M, Uhl J, Gutmann M and Leyk S. (2020) Change in U.S. small town community capitals, 1980–2010. *Population Research and Policy Review* 39: 913–940.

Ilbery B and Bowler I. (1998) From agricultural productivism to post-productivism. In: Ilbery B (ed.) *The Geography of Rural Change*. Abingdon: Pearson Education.

Ilbery B, Holloway L and Arber R. (1999) The geography of organic farming in England and Wales in the 1990s. *Tijdschrift voor economische en sociale geografie* 90: 285–295.

Ilbery B. (1978) Agricultural decision-making: A behavioural perspective. *Progress in Human Geography* 2: 448–466.

Ingold T. (2000) *The Perception of the Environment: Essays on Livelihood, Dwelling and Skill*. Hove: Psychology Press.

Ingold T. (2004) Culture on the ground: The world perceived through the feet. *Journal of Material Culture* 9: 315–340.

Ingold T. (2010) Footprints through the weather-world: Walking, breathing, knowing. *Journal of the Royal Anthropological Institute* 16: S121–S139.

Innes K. (2020) 'When I look at the landscape, I can read it' – practices of landscape interpretation in Detectorists. In: Innes K and Norcup J (eds) *Landscape of Detectorists*. Axminster: Uniform Books, 24–41.

Innes K and Norcup J. (2020) *Landscapes of Detectorists*. Axminister: Uniform Books.

Institute of Mechanical Engineers. (2013) *Global Food. Waste Not, Want Not*. London: Institute of Mechanical Engineers.

International Labour Office. (2017) *Global Estimates of Modern Slavery: Forced Labour and Forced Marriage*. Geneva: International Labour Office.

International Union for Conservation of Nature. (2020) *Protected Area Categories*. Available at: www.iucn.org/theme/protected-areas/about/protected-area-categories.

Inwood J and Bonds A. (2017) Property and whiteness: The Oregon standoff and the contradictions of the US settler state. *Space and Polity* 21: 253–268.

Ip M. (2003) Maori-Chinese encounters: Indigine-immigrant interaction in New Zealand. *Asian Studies Review* 27: 227–252.

IPCC. (2014) *Climate Change 2014: Synthesis Report. Contribution of Working Groups I, II and III to the Fifth Assessment Report of the Intergovernmental Panel on Climate Change* Geneva: IPCC.

IPPC. (2021) *Climate Change 2021: The Physical Science Basis. Summary for Policy Makers*. Cambridge: Cambridge University Press.

Jackson P. (1994) *Maps of Meaning: An Introduction to Cultural Geography*. Hove: Psychology Press.

Jacob S, Bourke L and Luloff A. (1997) Rural community stress, distress, and well-being in Pennsylvania. *Journal of Rural Studies* 13: 275–288.

Jakat, L. (2013) The Rosamunde Pilcher trail: Why German tourists flock to Cornwall. *The Guardian*. Available at: https://www.theguardian.com/travel/2013/oct/04/rosamunde-pilcher-german-tourists-cornwall

Jarosz L. (2008) The city in the country: Growing alternative food networks in Metropolitan areas. *Journal of Rural Studies* 24: 231–244.

Jenkins S. (2007) The assault on geography breeds ignorance and erodes nationhood. *The Guardian*. Available at: https://www.theguardian.com/commentisfree/2007/nov/16/comment.politics.

Johnson K and Lichter D. (2019) Rural depopulation: Growth and decline processes over the past century. *Rural Sociology* 84: 3–27.

Jones M, Jones R and Woods M. (2004) *An Introduction to Political Geography: Space, Place and Politics*. London; New York: Routledge.

Jones O. (1997) Little figures, big shadows: Country childhood stories. In: Cloke P and Little J (eds) *Contested Countryside Cultures: Otherness, Marginalisation and Rurality*. London: Routledge, 158–179.

Jones O. (2006) Non-human rural studies. In: Cloke P, Marsden T and Mooney P (eds) *Handbook of Rural Studies*. London: SAGE, 185–200.

Jones O and Cloke P. (2002) *Tree Cultures: The Place of Trees and Trees in Their Place*. Oxford: Berg Publisher.

Jones R and Tonts M. (1995) Rural restructuring and social sustainability: Some reflections on the Western Australian wheatbelt. *The Australian Geographer* 26: 133–140.

Jordan K, Krivokapic-Skoko B and Collins J. (2009) The ethnic landscape of rural Australia: Non-Anglo-Celtic immigrant communities and the built environment. *Journal of Rural Studies* 25: 376–385.

Jørgensen D. (2015) Rethinking rewilding. *Geoforum* 65: 482–488.

Juhola S, Klein N, Käyhkö J, and Schmid Neset T. (2017) Climate change transformations in Nordic agriculture? *Journal of Rural Studies* 51: 28–36.

Kallis G, Yarwood R and Tyrrell N. (2020) Gender, spatiality and motherhood: Intergenerational change in Greek-Cypriot migrant families in the UK. *Social & Cultural Geography* 1–18.

Kati V and Jari N. (2016) Bottom-up thinking – identifying socio-cultural values of ecosystem services in local blue – green infrastructure planning in Helsinki, Finland. *Land Use Policy* 50: 537–547.

Kearns A and Paddison R. (2000) New challenges for urban governance. *Urban Studies* 37: 845–850.

Kearns G and Philo C. (1993) *Selling Places: The City as Cultural Capital, Past and Present*. Oxford: Pergamon, 1–32.

Keeble D and Tyler P. (1995) Enterprising behaviour and the urban-rural shift. *Urban Studies* 32: 975–997.

Keeble D. (1980) Industrial decline, regional policy and the urban – rural manufacturing shift in the United Kingdom. *Environment and Planning A: Economy and Space* 12: 945–962.

Kelly C and Yarwood R. (2018) From rural citizenship to the rural citizen: Farming, dementia and networks of care. *Journal of Rural Studies* 63: 96–104.

Kinsman P. (1995) Landscape, race and national identity: The photography of Ingrid Pollard. *Area* 27: 300–310.

Kiryluk-Dryjska E, Beba P and Poczta W. (2020) Local determinants of the Common Agricultural Policy rural development funds' distribution in Poland and their spatial implications. *Journal of Rural Studies* 74: 201–209.

Kordel S and Pohle P. (2018) International lifestyle migration in the Andes of Ecuador: How migrants from the USA perform privilege, import rurality and evaluate their impact on local community. *Sociologia Ruralis* 58: 126–146.

Kortelainen J. (1999) The river as an actor-network: The Finnish forest industry utilization of lake and river systems. *Geoforum* 30: 235–247.

Kovács K. (2012) Rescuing a small village school in the context of rural change in Hungary. *Journal of Rural Studies* 28: 108–117.

Krawchenko T, Keefe J, Manuel P and Rapaport E. (2016) Coastal climate change, vulnerability and age friendly communities: Linking planning for climate change to the age friendly communities agenda. *Journal of Rural Studies* 44: 55–62.

Lake District National Park. (2022) *Tourists*. Available at: https://www.lakedistrict.gov.uk.

Laoire C. (2007) The 'green green grass of home'? Return migration to rural Ireland. *Journal of Rural Studies* 23: 332–344.

Laoire C and Stockdale A. (2016) Migration and the life course in rural settings. In: Shucksmith M and Brown M (eds) *Routledge International Handbook of Rural Studies*. London: Routledge, 36–49.

Lapping M and Scott M. (2019) The evolution of rural planning in the Global North. In: Scott M, Gallent N and Gkartios M (eds) *The Routledge Companion to Rural Planning*. London: Routledge, 28–45.

Lavau S. (2017) Viruses. In: Adey P, Bissell D, Hannam K, Merriman P and Sheller M. (eds) *The Routledge Handbook of Mobilities*. London: Routledge.

La Via Campesina. (2022) *La Via Campesina's International Campaign*. Available at: https://viacampesina.org/en/.

Laviolette P. (2016) *Extreme Landscapes of Leisure: Not a Hap-Hazardous Sport*. London: Routledge.

Law A, Gaywood M, Jones K, Ramsay P and Willby N. (2017) Using ecosystem engineers as tools in habitat restoration and rewilding: Beaver and wetlands. *Science of the Total Environment* 605–606: 1021–1030.

Law R. (1999) Beyond 'women and transport': Towards new geographies of gender and daily mobility. *Progress in Human Geography* 23: 567–588.

Lawrence G. (2016) Food systems and land: Connections and contradictions. In: Shucksmith M and Brown D (eds) *Routledge International Handbook of Rural Studies*. London: Routledge, 183–191.

Lawrence M. (1995) Rural homelessness: A geography without a geography. *Journal of Rural Studies* 11: 297–307.

Lawson V. (2007) Geographies of care and responsibility. *Annals of the Association of American Geographers* 97: 1–11.

Lay J, Ward A, Eckery A, et al. (2021) *Taking Stock of the Global Land Rush: Few Development Benefits, Many Human and Environmental Risks. Analytical Report III*. Bern; Montpellier; Hamburg; Pretoria: Land Matrix.

Leach B. (2016) Jobs for women? Gender and class in Ontario's ruralized Automotive Manufacturing Industry. In: Pini B and Leach B (eds) *Reshaping Gender and Class in Rural Spaces*. London: Routledge, 129–144.

Leckie G. (1996) 'They never trusted me to drive': Farm girls and the gender relations of agricultural information transfer. *Gender, Place & Culture* 3: 309–326.

Lee J. (2007) Experiencing landscape: Orkney hill land and farming. *Journal of Rural Studies* 23: 88–100.

Lefebvre H. (1974) *The Production of Space*. Oxford: Blackwell.

Leslie D. (1993) Femininity, post-Fordism, and the 'new traditionalism'. *Environment and Planning D: Society and Space* 11: 689–708.

Lewis G. (1998) Rural migration and demographic change. In: Ilbery B (ed.) *The Geography of Rural Change*. London: Routledge, 131–160.

Ley D. (1987) Reply: The rent gap revisited. *Annals of the Association of American Geographers* 77: 465–468.

Leyshon M. (2002) On being 'in the field': Practice, progress and problems in research with young people in rural areas. *Journal of Rural Studies* 18: 179–191.

Leyshon M. (2008a) The betweeness of being a rural youth: Inclusive and exclusive lifestyles. *Social & Cultural Geography* 9: 1–26.

Leyshon M. (2008b) 'We're stuck in the corner': Young women, embodiment and drinking in the countryside. *Drugs: Education, Prevention and Policy* 15: 267–289.

Liepins R. (2000a) Exploring rurality through 'community': Discourses, practices and spaces shaping Australian and New Zealand rural 'communities'. *Journal of Rural Studies* 16: 325–341.

Liepins R. (2000b) New energies for an old idea: Reworking approaches to 'community' in contemporary rural studies. *Journal of Rural Studies* 16: 23–35.

Lillemets J, Fertő I and Viira A. (2022) The socioeconomic impacts of the CAP: Systematic literature review. *Land Use Policy* 114: 105968.

Lippe R, Cui S and Schweinle J. (2021) Estimating global forest-based employment. *Forests* 12: 1219.

Little J. (1987) Gender relations in rural-areas – the importance of women's domestic role. *Journal of Rural Studies* 3: 335–342.

Little J. (1991) Theoretical issues of women's non-agricultural employment in rural areas, with illustrations from the U.K. *Journal of Rural Studies* 7: 99–105.

Little J. (2002a) *Gender and Rural Geography: Identity, Sexuality and Power in the Countryside*. London: Routledge.

Little J. (2002b) Rural geography: Rural gender identity and the performance of masculinity and femininity in the countryside. *Progress in Human Geography* 26: 665–670.

Little J. (2007) Constructing nature in the performance of rural heterosexualities. *Environment and Planning D: Society and Space* 25: 851–866.

Little J. (2016) Gender and entrepreneurship. In: Shucksmith M and Brown D (eds) *Routledge International Handbook of Rural Studies*. London: Routledge, 434–445.

Little J. (2017) Understanding domestic violence in rural spaces: A research agenda. *Progress in Human Geography* 41: 472–488.

Little J and Austin P. (1996) Women and the rural idyll. *Journal of Rural Studies* 12: 101–111.

Little J and Jones O. (2000) Masculinity, gender, and rural policy. *Rural Sociology* 65: 621–639.

Little J and Leyshon M. (2003) Embodied rural geographies: Developing research agendas. *Progress in Human Geography* 27: 257–272.

Little J and Morris C. (2005) *Critical Studies in Rural Gender Issues*. Aldershot: Ashgate.

Little J, Panelli R and Kraack A. (2005) Women's fear of crime: A rural perspective. *Journal of Rural Studies* 21: 151–163.

Little J and Panelli R. (2003) Gender research in rural geography. *Gender, Place & Culture* 10: 281–289.

Liu Y, Liu J and Zhou Y. (2017) Spatio-temporal patterns of rural poverty in China and targeted poverty alleviation strategies. *Journal of Rural Studies* 52: 66–75.

Lockie S, Lawrence G and Cheshire L. (2006) Reconfiguring rural resource governance: The legacy of neo-liberalism in Australia. In: Cloke P, Goodwin M and Mooney P (eds) *Handbook of Rural Studies*. London: SAGE, 29–43.

LOCOG (London Organising Committee of the Olympic Games and Paralympic Games).2012. London 2012 Olympic Games Opening Ceremony Media Guide. London:London Organising Committee of the Olympic Games and Paralympic Games. http://www.webarchive.nationalarchives.gov.uk/20120730004223/http://www.london2012.com/mm/Document/Documents/Publications/01/30/43/40/OPENINGCEREMONYGUIDE_English.pdf Accessed August 27, 2013.

London National Park City. (2022) *Welcome to London National Park City*. Available at: https://www.nationalparkcity.london/.

Longhurst R. (1995) VIEWPOINT the body and geography. *Gender, Place & Culture* 2: 97–106.

Lord Taylor of Goss Moor, Essex S and Wilson O. (2022) Solving the housing market crisis in England and Wales: From new towns to garden communities. *Geography* 107: 4–13.

Lorimer H and Lund K. (2003) Performing facts: Finding a way over Scotland's mountains. *The Sociological Review* 51: 130–144.

Lovett A, Haynes R, Sünnenberg G, and Gale S. (2002) Car travel time and accessibility by bus to general practitioner services: A study using patient registers and GIS. *Social Science & Medicine* 55: 97–111.

Low S, Bass M, Thilmany D, and Castillo M. (2021) Local foods go downstream: Exploring the spatial factors driving U.S. Food Manufacturing. *Applied Economic Perspectives and Policy* 43: 896–915.

Lowe P and Ward N. (2001) New labour, new rural vision? Labour's rural white paper. *The Political Quarterly* 72: 386–390.

Lowe P and Ward N. (2009) England's rural futures: A socio-geographical approach to scenarios analysis. *Regional Studies* 43: 1319–1332.

MacEwen A and MacEwen M. (1987) *Greenprints for the Countryside?* Hemel Hempstead: Allen & Unwin.

Macfarlane R. (2012) *The Old Ways: A Journey on Foot*. London: Penguin.

MacKian S. (1995) 'That great dust-heap called history':* Recovering the multiple spaces of citizenship. *Political Geography* 14: 209–216.

MacKrell P and Pemberton S. (2018) New representations of rural space: Eastern European migrants and the denial of poverty and deprivation in the English countryside. *Journal of Rural Studies* 59: 49–57.

Macnaghten P and Urry J. (1998) *Contested Natures*. London; Thousand Oaks, CA: SAGE.

Macpherson H. (2008) "I don't know why they call it the Lake District they might as well call it the rock district!" The workings of humour and laughter in research with members of visually impaired walking groups. *Environment and Planning D: Society & Space* 26: 1080–1095.

Macpherson H. (2009) The intercorporeal emergence of landscape: Negotiating sight, blindness, and ideas of landscape in the British countryside. *Environment and Planning A: Economy and Space* 41: 1042–1054.

Macpherson H. (2010) Non-representational approaches to body – landscape relations. *Geography Compass* 4: 1–13.

Macrotrends (2023) *Wheat prices: 40 year historical chart*. http://www.macrotrends.net/2534/wheat-prices-historical-chart-data

MacTavish K, Eley M and Salamon S. (2006) Housing vulnerability among rural trailer-park households. *Georgetown Journal on Poverty Law & Policy* 13: 95.

MacTavish K. (2020) Coming of age on the Edge of Town: Perspectives in growing up in a rural trailer park. In: Burrow A and Hill P (eds) *The Ecology of Purposeful Living across the Lifespan*. New York: Springer, 137–148.

Magnusson L and Turner B. (2008) Municipal housing companies in Sweden – social by default. *Housing, Theory and Society* 25: 275–296.

Maher G and Cawley M. (2016) Short-term labour migration: Brazilian migrants in Ireland. *Population, Space and Place* 22: 23–35.

Mahon M and Hyyryläinen T. (2019) Rural arts festivals as contributors to rural development and resilience. *Sociologia Ruralis* 59: 612–635.

Mansvelt J. (2005) *Geographies of Consumption*. London: SAGE.

Marshall T. (1950) *Social Class and Citizenship*. London: Hutchinson.

Marston S, Jones J III and Woodward K. (2005) Human geography without scale. *Transactions of the Institute of British Geographers* 30: 416–432.

Massey D and Allen J. (1984) *Geography Matters*. Milton Keynes: Open University.

Massey D. (1991) A global sense of place. *Marxism Today* 38: 24–29.

Massey D. (1994) *Space, Place, and Gender*. Minneapolis, MN: University of Minnesota Press.

Massey D. (2005) *For Space*. London; Thousand Oaks, CA: SAGE.

Matarrita-Cascante D, Sene-Harper A and Stocks G. (2015) International amenity migration: Examining environmental behaviors and influences of amenity migrants and local residents in a rural community. *Journal of Rural Studies* 38: 1–11.

Mather A, Hill G, and Nijnik M. (2006) Post-productivism and rural land use: Cul de sac or challenge for theorization? *Journal of Rural Studies* 22: 441–455.

Matless D. (2016) *Landscape and Englishness: Second Expanded Edition*. London: Reaktion Books.

Matthews H, Taylor M, Sherwood K, Tucker F and Limb M (2000) Growing-up in the countryside: Children and the rural idyll. *Journal of Rural Studies* 16: 141–153.

Matthews P, Bramley G and Hastings A. (2015) Homo economicus in a big society: Understanding middle-class Activism and NIMBYism towards new housing developments. *Housing, Theory and Society* 32: 54–72.

Mawby R and Yarwood R. (2010) *Rural Policing and Policing the Rural: A Constable Countryside?* London: Routledge.

Maxey L. (2006) Can we sustain sustainable agriculture? Learning from small-scale producer-suppliers in Canada and the UK. *Geographical Journal* 172: 230–244.

Maye D. (2016) Geographies of food production. In: Daniels P, Bradshaw M, Shaw D, Sidaway J and Hall T. (eds) *An Introduction to Human Geography*. London: Pearson Education.

Mayer A, Olson-Hazboun S and Malin S. (2018) Fracking fortunes: Economic well-being and oil and gas development along the urban-rural continuum. *Rural Sociology* 83: 532–567.

Mbow C, Rosenzweig C, Barioni L, et al. (2019) Food security. In: IPPC (ed.) *Climate Change and Land*. Geneva; Rome: IPCC, 437–550.

McAreavey R. (2016) Understanding the association between rural ethnicity and inequalities. In: Shuksmith M and Brown D (eds) *Routledge International Handbook of Rural Studies*. London: Routledge, 477–494.

McCarthy J. (2006a) Neoliberalism and the politics of alternatives: Community forestry in British Columbia and the United States. *Annals of the Association of American Geographers* 96: 84–104.

McCarthy J. (2006b) Rural geography: Alternative rural economies – the search for alterity in forests, fisheries, food, and fair trade. *Progress in Human Geography* 30: 803–811.

McDonagh J. (2014) Rural geography II: Discourses of food and sustainable rural futures. *Progress in Human Geography* 38: 838–844.

McDowell L. (1983) Towards an understanding of the gender division of urban space. *Environment and Planning D: Society & Space* 1: 59–72.

McDowell L. (1992) Doing gender – feminism, feminists and research methods in human-geography. *Transactions of the Institute of British Geographers* 17: 399–416.

McDowell L. (1994) The transformation of cultural geography. In: Gregory D, Martin R and Smith G (eds) *Human Geography: Society, Space and Social Science*. London: Springer, 146–173.

McEntee J and Agyeman J. (2010) Towards the development of a GIS method for identifying rural food deserts: Geographic access in Vermont, USA. *Applied Geography* 30: 165–176.

McEwen L, Hall T, Hunt J, Dempsey M and Harrison M. (2002) Flood warning, warning response and planning control issues associated with caravan parks: The April 1998 floods on the lower Avon floodplain, Midlands region, UK. *Applied Geography* 22: 271–305.

McGee T and Russell S. (2003) "It's just a natural way of life . . ." an investigation of wildfire preparedness in rural Australia. *Global Environmental Change Part B: Environmental Hazards* 5: 1–12.

McGrath E, Harmer N and Yarwood R. (2020) Ferries as travelling landscapes: Tourism and watery mobilities. *International Journal of Culture, Tourism and Hospitality Research* 321–334.

McKenzie F, Haslam McKenzie F and Hoath A. (2014) Fly-in/fly-out, flexibility and the future: Does becoming a regional FIFO source community present opportunity or burden? *Geographical Research* 52: 430–441.

McKenzie F and Hoath A. (2014) The socio-economic impact of mine industry commuting labour force on source communities. *Resources Policy* 42: 45–52.

McLaughlin B. (1986a) Rural policy in the 1980s: The revival of the rural idyll. *Journal of Rural Studies* 2: 81–90.

McLaughlin B. (1986b) The rhetoric and the reality of rural deprivation. *Journal of Rural Studies* 2: 291–307.

McLaughlin B. (1987) Rural policy into the 1990s – self help or self deception. *Journal of Rural Studies* 3: 361–364.

McManus P, Walmsley J, Argent N, and Baum S, Bourke L, Martin J, Pritchard B and Sorensen T. (2012) Rural community and rural resilience: What is important to farmers in keeping their country towns alive? *Journal of Rural Studies* 28: 20–29.

McMillan Lequieu A. (2017) "We made the choice to stick it out": Negotiating a stable home in the rural, American Rust Belt. *Journal of Rural Studies* 53: 202–213.

Medway D and Warnaby G. (2014) What's in a name? Place branding and toponymic commodification. *Environment and Planning A: Economy and Space* 46: 153–167.

Meijering L, Lettinga A, Nanninga C, and Milligan C. (2017) Interpreting therapeutic landscape experiences through rural stroke survivors' biographies of disruption and flow. *Journal of Rural Studies* 51: 275–283.

Mercer I. (2002) *Crisis and Opportunity: Devon Foot and Mouth Inquiry 2001.* Tiverton: Devon Books.

Meyer K and Lobao L. (2003) Economic hardship, religion and mental health during the midwestern farm crisis. *Journal of Rural Studies* 19: 139–155.

Meyer M. (2019) Climate change, environmental hazards and community sustainability. In: Shucksmith M and Brown D. (eds) *Routledge International Handbook of Rural Studies.* London: Routledge.

Michel-Villarreal R, Hingley M, Canavari M, and Bregoli, I. (2019) Sustainability in alternative food networks: A systematic literature review. *Sustainability* 11: 859.

Middeldorp N and Le Billon P. (2019) Deadly environmental governance: Authoritarianism, eco-populism, and the repression of environmental and land defenders. *Annals of the American Association of Geographers* 109: 324–337.

Miele M and Lever J. (2013) Civilizing the market for welfare friendly products in Europe? The techno-ethics of the Welfare Quality® assessment. *Geoforum* 48: 63–72.

Milbourne P. (1997) *Revealing Rural 'Others': Representation, Power, and Identity in the British Countryside.* London; Washington, DC: Pinter.

Milbourne P. (2004) The local geographies of poverty: A rural case-study. *Geoforum* 35: 559–575.

Milbourne P. (2006) Rural housing and homelessness. In: Cloke P, Goodwin M and Mooney P (eds) *Handbook of Rural Studies.* London: SAGE, 427–444.

Milbourne P. (2007) Re-populating rural studies: Migrations, movements and mobilities. *Journal of Rural Studies* 23: 381–386.

Milbourne P. (2014) Poverty, place, and rurality: Material and sociocultural disconnections. *Environment and Planning A* 46: 566–580.

Milbourne P. (2016) Poverty and welfare in rural places. In: Shucksmith M and Brown D (eds) *Routledge International Handbook of Rural Studies.* London: Routledge, 480–491.

Milbourne P. (2017) Rural geography. *International Encyclopedia of Geography* 1–17.

Milbourne P and Doheny S. (2012) Older people and poverty in rural Britain: Material hardships, cultural denials and social inclusions. *Journal of Rural Studies* 28: 389–397.

Milbourne P and Kitchen L. (2014) Rural mobilities: Connecting movement and fixity in rural places. *Journal of Rural Studies* 34: 326–336.

Miller S. (1996a) Class, power and social construction: Issues of theory and application in thirty years of rural studies. *Sociologia Ruralis* 36: 93.

Miller S. (1996b) Theory, application and critical practice: Rejoinder. *Sociologia Ruralis* 36: 365–370.

Mingay G. (1989) *The Rural Idyll*. London: Routledge.

Mitchell C. (2004) Making sense of counter urbanization. *Journal of Rural Studies* 20: 15–34.

Mitchell D. (2000) *Cultural Geography: A Critical Introduction*. Oxford: Blackwell.

Mittal A. (2009) *The 2008 Food Price Crisis: Rethinking Food Security Policies*. New York: UN.

Mohan G and Power M. (2009) Africa, China and the 'new' economic geography of development. *Singapore Journal of Tropical Geography* 30: 24–28.

Monbiot, G (2009) *Our Once and Future World* https://www.monbiot.com/2019/09/27/our-once-and-future-world/

Monbiot G. (2013) *Sheepwrecked: How Britain Has Been Shagged by the White Plague*. Available at: www.monbiot.com/2013/05/30/sheepwrecked/.

Monbiot G. (2014) *Feral: Rewilding the Land, the Sea, and Human Life*. Chicago, IL: University of Chicago Press.

Mormont M. (1987) Rural nature and urban natures. *Sociologia Ruralis* 27: 1–20.

Mormont M. (1990) Who is rural? or, how to be rural: Towards a sociology of the rural. In: Marsden T, Whatmore S and Lowe P (eds) *Rural Restructuring: Global Processes and Their Responses*. London: David Fulton Publishers, 21–44.

Morris C and Buller H. (2003) The local food sector. *British Food Journal* 105: 559–566.

Morris C and Evans N. (2001) 'Cheese makers are always women': Gendered representations of farm life in the agricultural press. *Gender, Place & Culture* 8: 375–390.

Morris C and Evans N. (2004) Agricultural turns, geographical turns: Retrospect and prospect. *Journal of Rural Studies* 20: 95–111.

Morris C and Young C. (1997) Towards environmentally beneficial farming? An evaluation of the Countryside Stewardship Scheme. *Geography* 305–316.

Moseley M. (1979) *Accessibility: The Rural Challenge*. London: Methuen and Co.

Moseley M. (1980) Is rural deprivation really rural? *The Planner* 66: 97.

Moseley M. (1997) Parish appraisals as a tool of rural community development: An assessment of the British experience. *Planning Practice & Research* 12: 197–212.

Moseley M. (2003) *Rural Development: Principles and Practice*. New York: SAGE.

Mouat M and Prince R. (2018) Cultured meat and cowless milk: On making markets for animal-free food. *Journal of Cultural Economy* 11: 315–329.

Mueller J and Tickamyer A. (2020) Climate change beliefs and support for development: Testing a cognitive hierarchy of support for natural resource-related economic development in rural Pennsylvania. *Journal of Rural Studies* 80: 553–566.

Muir R. (2012) *Pubs and Places: The Social Value of Community Pubs*. London: Institute for Public Policy Research.

Müller D. (2013) Progressing second home research: A Nordic perspective. *Scandinavian Journal of Hospitality and Tourism* 13: 273–280.

Munday M, Bristow G and Cowell R. (2011) Wind farms in rural areas: How far do community benefits from wind farms represent a local economic development opportunity? *Journal of Rural Studies* 27: 1–12.

Murdoch J. (1997) The shifting territory of government: Some insights from the Rural White Paper. *Area* 29: 109–118.

Murdoch J. (2000) Networks – a new paradigm of rural development? *Journal of Rural Studies* 16: 407–419.

Murdoch J. (2003) Co-constructing the countryside: Hybrid networks and the extensive self. *Country Visions* 263–282.

Murdoch J, Lowe P and Ward N, (2003) *The Differentiated Countryside*. London; New York: Routledge.

Murdoch J and Pratt A. (1993) Rural studies – modernism, postmodernism and the post rural. *Journal of Rural Studies* 9: 411–427.

Murdoch J and Pratt A. (1994) Rural studies of power and the power of rural studies – a reply. *Journal of Rural Studies* 10: 83–87.

Murdoch, J (2006) Networking rurality: emergent complexity in the countryside in Cloke, P, Marsden, T and Mooney, P The Handbook of Rural Studies Sage, London, 171-184

Nacu A. (2011) The politics of Roma migration: Framing identity struggles among Romanian and Bulgarian Roma in the Paris region. *Journal of Ethnic and Migration Studies* 37: 135–150.

Nadin V, Hart T, Davoudi S, Webb D, Vigar G, Pendlebury J and Townshend T. (2014) *Town and Country Planning in the UK*. London: Routledge.

Nash R. (1970) The American invention of national parks. *American Quarterly* 22: 726–735.

Natale F, Kalantaryan S, Scipioni M, Alessandrini A and Pasa, A. (2019) *Migration in EU Rural Areas*. Luxembourg: Publications Office of the European Union.

National Park Service. (2020) *Origins of National Parks*. Available at: www.nps.gov/articles/npshistory-origins.htm.

National Rural Health Alliance. (2017) *Poverty in Rural and Remote Australia*. Deakin West, ACT: National Rural Health Alliance.

National Union of Farmers. (2022) *Black British Farming*. Available at: www.nfuonline.com/sectors/student-farmer/black-british-farming.

Naumann M and Rudolph D. (2020) Conceptualizing rural energy transitions: Energizing rural studies, ruralizing energy research. *Journal of Rural Studies* 73: 97–104.

Neal S. (2009) *Rural Identities: Ethnicity and Community in the Contemporary English Countryside*. Farnham: Ashgate.

Neal S and Agyeman J. (2006) *The New Countryside? Ethnicity, Nation and Exclusion in Contemporary Rural Britain*. Bristol: Policy Press.

Neal S and Walters S. (2006) Strangers asking strange questions? A methodological narrative of researching belonging and identity in English rural communities. *Journal of Rural Studies* 22: 177–189.

Nelson L and Nelson P. (2011) The global rural: Gentrification and linked migration in the rural USA. *Progress in Human Geography* 35: 441–459.

Newby H. (1979a) *The Deferential Worker: A Study of Farm Workers in East Anglia*. Madison, WI: University of Wisconsin Press.

Newby H. (1979b) *Green and Pleasant Land? Social Change in Rural England*. London: Hutchinson of London.

Newby H. (1986) Locality and rurality: The restructuring of rural social relations. *Regional Studies* 20: 209–215.

Newby H, Bell C, Rose D, and Saunders P. (1978) *Property, Paternalism and Power*. London: Hutchinson.

Ni Laoire C. (2002) Young farmers, masculinities and change in rural Ireland. *Irish Geography* 35: 16–27.

Nogué J and Vicente J. (2004) Landscape and national identity in Catalonia. *Political Geography* 23: 113–132.

Nolte K, Chamberlain W and Giger M. (2016) *International Land Deals for Agriculture, Fresh Insights from the Land Matrix: Analytical Report II*. Pretoria: Landmatrix.

Norris A, Zajicek A and Murphy-Erby Y. (2010) Intersectional perspective and rural poverty research: Benefits, challenges and policy implications. *Journal of Poverty* 14: 55–75.

Notter M, MacTavish K and Shamah D. (2008) Pathways toward resilience among women in rural trailer parks. *Family Relations* 57: 613–624.

Office for National Statistics. (2016) *2011 Rural/Urban Classification*. Available at: www.ons.gov.uk/methodology/geography/geographicalproducts/ruralurbanclassifications/2011ruralurbanclassification.

Office of National Statistics (2012) Second Home Ownership in England and Wales https://www.ons.gov.uk/peoplepopulationandcommunity/housing/bulletins/2011censusnu Accessed 21/01/2021

Office for National Statistics. (2019) *Housing Affordability in England and Wales: 2019*. London: Office for National Statistics.

Oglethorpe D. (2009) Food miles – the economic, environmental and social significance of the focus on local food. *CAB Reviews: Perspectives in Agriculture, Veterinary Science, Nutrition and Natural Resources* 4: 1–11.

O'Halloran M and Davies A. (2020) A shared risk: Volunteer shortages in Australia's rural bushfire brigades. *Australian Geographer* 51: 421–435.

Olsson L, Jerneck A, Thoren H, Persson J and O'Byrne D. (2015) Why resilience is unappealing to social science: Theoretical and empirical investigations of the scientific use of resilience. *Science Advances* 1: E1400217.

Orange, P (2023) The English Countryside: life in an English Village https://www.expatica.com/uk/moving/location/english-countryside-103181/Accessed 09/02/2023

Ottelin J, Heinonen J, Nässén J, Junnila S. (2019) Household carbon footprint patterns by the degree of urbanisation in Europe. *Environmental Research Letters* 14: 114016.

Overvåg K and Berg N. (2011) Second homes, rurality and contested space in Eastern Norway. *Tourism Geographies* 13: 417–442.

Owen S and Carrington K. (2015) Domestic violence (DV) service provision and the architecture of rural life: An Australian case study. *Journal of Rural Studies* 39: 229–238.

Owen S and Moseley M. (2003) Putting Parish Plans in their place: Relationships between community-based initiatives and development planning in English villages. *The Town Planning Review* 445–471.

Pacione M. (1995) The geography of deprivation in rural Scotland. *Transactions of the Institute of British Geographers* 20: 173–192.

Page B and Walker R. (1991) From settlement to Fordism: The agro-industrial revolution in the American Midwest. *Economic Geography* 67: 281–315.

Pahl R. (1965a) Class and community in English commuter villages. *Sociologia Ruralis* 5: 5–23.

Pahl R. (1965b) *Urbs in Rure; The Metropolitan Fringe in Hertfordshire*. London: LSE.

Pain R. (2014) Everyday terrorism: Connecting domestic violence and global terrorism. *Progress in Human Geography* 38: 531–550.

Palmer C and Brady E. (2007) Landscape and value in the work of Alfred Wainwright (1907–1991). *Landscape Research* 32: 397–421.

Panelli R, Gallagher L and Kearns R. (2006) Access to rural health services: Research as community action and policy critique. *Social Science & Medicine* 62: 1103–1114.

Panelli R, Hubbard P, Coombes B, and Suchet-Pearson S. (2009) De-centring White ruralities: Ethnic diversity, racialisation and indigenous countrysides. *Journal of Rural Studies* 25: 355–364.

Panelli R, Little J and Kraack A. (2004) A community issue? Rural women's feelings of safety and fear in New Zealand. *Gender, Place & Culture* 11: 445–467.

Panelli R, Nairn K and McCormack J. (2002) "We make our own fun": Reading the politics of youth with(in) community. *Sociologia Ruralis* 42: 106–130.

Park S. (2017) Digital inequalities in rural Australia: A double jeopardy of remoteness and social exclusion. *Journal of Rural Studies* 54: 399–407.

Parker G. (1999) The role of the consumer-citizen in environmental protest in the 1990s. *Space and Polity* 3: 67–83.

Parker G. (2006) The Country Code and the ordering of countryside citizenship. *Journal of Rural Studies* 22: 1–16.

Parker G, Lynn T and Wargent M. (2015) Sticking to the script? The co-production of neighbourhood planning in England. *Town Planning Review* 86: 519–536.

Parr H and Philo C. (2003) Rural mental health and social geographies of caring. *Social & Cultural Geography* 4: 471–488.

Parr H, Philo C and Burns N. (2004) Social geographies of rural mental health: Experiencing inclusions and exclusions. *Transactions of the Institute of British Geographers* 29: 401–419.

Pattison G. (2004) Planning for decline: The 'D'-village policy of County Durham, UK. *Planning Perspectives* 19: 311–332.

Peacock A and Pemberton S. (2019) The paradox of mobility for older people in the rural-urban fringe. *Journal of Rural Studies* 70: 9–18.

Pearce F. (2012) *The Land Grabbers: The New Fight over Who Owns the Earth*. Boston: Beacon Press.

Peeren E and Souch I. (2019) Romance in the cowshed: Challenging and reaffirming the rural idyll in the Dutch reality TV show farmer wants a wife. *Journal of Rural Studies* 67: 37–45.

Pemberton S. (2019) *Rural Regeneration in the UK*. London: Routledge.

Periäinen K. (2006) The summer cottage: A dream in the Finnish forest. In: McIntyre N, Williams D and McHugh K (eds) *Multiple Dwelling and Tourism*. Cambridge, MA: CABI, 103–113.

Perkins H, Mackay M and Espiner S. (2015) Putting pinot alongside merino in Cromwell District, Central Otago, New Zealand: Rural amenity and the making of the global countryside. *Journal of Rural Studies* 39: 85–98.

Perrings C. (1998) Introduction: Resilience and sustainability. *Environment and Development Economics* 3: 221–262.

Persson L. (2019) Lifestyle migrants or "environmental refugees"? – Resisting urban risks. *Population, Space and Place* 25: E2254.

Peter G, Bell M, Jarnagin S and Bauer D. (2000) Coming back across the fence: Masculinity and the transition to sustainable agriculture. *Rural Sociology* 65: 215–233.

Phillips M. (1993) Rural gentrification and the processes of class colonization. *Journal of Rural Studies* 9: 123–140.

Phillips M. (2002) The production, symbolization and socialization of gentrification: Impressions from two Berkshire villages. *Transactions of the Institute of British Geographers* 27: 282–308.

Phillips M. (2014) Baroque rurality in an English village. *Journal of Rural Studies* 33: 56–70.

Phillips M, Fish R and Agg J. (2001) Putting together ruralities: Towards a symbolic analysis of rurality in the British mass media. *Journal of Rural Studies* 17: 1–27.

Phillips M and Smith D. (2018a) Comparative approaches to gentrification: Lessons from the rural. *Dialogues in Human Geography* 8: 3–25.

Phillips M and Smith D. (2018b) Comparative ruralism and "opening new windows' on gentrification. *Dialogues in Human Geography* 8: 51–58.

Philo C. (1992) Neglected rural geographies: A review. *Journal of Rural Studies* 8: 193–207.

Philo C. (1993) Postmodern rural geography – a reply to Murdoch and Pratt. *Journal of Rural Studies* 9: 429–436.

Philo C. (1995) Animals, geography, and the city: Notes on inclusions and exclusions. *Environment and Planning D: Society and Space* 13: 655–681.

Philo C, Parr H and Burns N. (2003) Rural madness: A geographical reading and critique of the rural mental health literature. *Journal of Rural Studies* 19: 259–281.

Philo C, Parr H and Burns N. (2017) The rural panopticon. *Journal of Rural Studies* 51: 230–239.

Philo C and Wilbert C. (2000) *Animal Spaces, Beastly Places: New Geographies of Human-Animal Relations*. London; New York: Routledge.

Pini B. (2002) Focus groups, feminist research and farm women: Opportunities for empowerment in rural social research. *Journal of Rural Studies* 18: 339–351.

Pini B. (2004) On being a nice country girl and an academic feminist: Using reflexivity in rural social research. *Journal of Rural Studies* 20: 169–179.

Pini B. (2006) A critique of 'new' rural local governance: The case of gender in a rural Australian setting. *Journal of Rural Studies* 22: 396–408.

Pini B, Mayes R and Castro L. (2020) Rurality, geography and feminism: Troubling relationships. In: Datte A, Hopkins P, Johnston L, Olson E and Silva J. (eds) *Routledge Handbook of Gender and Feminist Geographies*. London: Routledge, 202–211.

Pini B, Philo C and Chouinard V. (2017) On making disability in rural places more visible: Challenges and opportunities [Introduction to a special issue]. *Journal of Rural Studies* 51: 223–229.

Poore J and Nemecek T. (2018) Reducing food's environmental impacts through producers and consumers. *Science* 360: 987–992.

Power E. (2005) Human – nature relations in suburban gardens. *Australian Geographer* 36: 39–53.

Power M, Mohan G and Tan-Mullins M. (2012) *China's Resource Diplomacy in Africa: Powering Development?* London: Palgrave Macmillan.

Pratt A. (1992) Book review, Laura Ashley: A life by design. *Journal of Rural Studies* 8: 126–127.

Pratt A. (1996) Discourses of rurality: Loose talk or social struggle? *Journal of Rural Studies* 12: 69–78.

Price L and Evans N. (2006) From 'as good as gold' to 'gold diggers': Farming women and the survival of British family farming. *Sociologia Ruralis* 46: 280–298.

Price L and Evans N. (2009) From stress to distress: Conceptualizing the British family farming patriarchal way of life. *Journal of Rural Studies* 25: 1–11.

Price L and Hawkins H. (2018) *Geographies of Making, Craft and Creativity*. London: Routledge.

Publications Office of the EU. (2020) *LEADER Local Action Groups Map*. Available at: https://op.europa.eu/en/publication-detail/-/publication/d186a784-30e6-43ed-baa6-30ca4f3ba5b8

Putnam R. (2000) Bowling alone: America's declining social capital. In: *Culture and Politics*. New York: Springer.

Quixote, D (2011) Welcome to the Highlands https://windfarmaction.wordpress.com/ Accessed 17/02/2023

Rainer G. (2019) Amenity/lifestyle migration to the Global South: Driving forces and socio-spatial implications in Latin America. *Third World Quarterly* 40: 1359–1377.

Rannikko P and Salmi P. (2018) Towards neo-productivism? – Finnish paths in the use of forest and sea. *Sociologia Ruralis* 58: 625–643.

Ray C. (2000) The EU LEADER programme: Rural development laboratory. *Sociologia Ruralis* 40: 163–171.

Rees A. (1950) *Life in a Welsh Countryside: A Social Study of Llanfihangel yng Ngwynfa*. Cardiff: University of Wales Press.

Relph E. (1981) *1981: Rational Landscapes and Humanistic Geography*. London: Croom Helm.

Ribchester C and Edwards B. (1999) The centre and the local: Policy and practice in rural education provision. *Journal of Rural Studies* 15: 49–63.

Rignall K and Atia M. (2017) The global rural: Relational geographies of poverty and uneven development. *Geography Compass* 11: E12322.

Riley M. (2009) 'The next link in the chain': Children, agri-cultural practices and the family farm. *Children's Geographies* 7: 245–260.

Riley M. (2018) Merging masculinities: Exploring intersecting masculine identities on family farms. In: Shorthall S and Bock B (eds) *Gender and Rural Globalisation: International Perspectives on Gender and Rural Development*. London: CAB International.

Ringwood K. (2013) *Dartmoor's Tors and Rocks*. Plymouth: University of Plymouth Press.

Riseth J. (2007) An indigenous perspective on national parks and Sami reindeer management in Norway. *Geographical Research* 45: 177–185.

Robinson E. (2021) Citizens, custodians, and villains: Environmentality and the politics of difference in Senegal's community forests. *Geoforum* 125: 25–36.

Robinson G. (1990) *Conflict and Change in the Countryside: Rural Society, Economy, and Planning in the Developed World*. London; New York: Belhaven Press.

Robinson G. (2004) *Geographies of Agriculture*. Harlow: Prentice Hall.

Robinson G. (2009) Towards sustainable agriculture: Current debates. *Geography Compass* 3: 1757–1773.

Robinson G. (2018) Agricultural geography. In: Richardson D, Castree N, Goodchild M, Kobayashi A, Liu W and Marston R (eds) *International Encyclopedia of Geography*. New York: Wiley Online, 1–15.

Robinson G. (2019) Sustainability and resilient ruralities. In: Scott M, Gallent N and Gkartzois M (eds) *The Routledge Companion to Rural Planning*. London: Routledge, 56–68.

Robson E. (2004) Children at work in rural northern Nigeria: Patterns of age, space and gender. *Journal of Rural Studies* 20: 193–210.

Rogers B. (2019) "Contrary to all the other shit I've said": Trans men passing in the south. *Qualitative Sociology* 42: 639–662.

Rose D. (1984) Rethinking gentrification: Beyond the uneven development of Marxist urban theory. *Environment and Planning D: Society and Space* 2: 47–74.

Rose N. (1996) Governing advanced liberal democracies. In: Barry A, Osborne T and Rose N (eds) *Foucault and Political Reason: Liberalism, Neo-liberalism and the Rationalities of Government*. Chicago, IL University of Chicago Press, 37–64.

Rosenfield S and Wojan T. (2018) The emerging contours of rural manufacturing. In: Shucksmith M and Brown D (eds) *The Routledge International Handbook of Rural Studies*. London: Routledge, 120–132.

Rosin C. (2013) Food security and the justification of productivism in New Zealand. *Journal of Rural Studies* 29: 50–58.

Rowe F and Patias N. (2020) Mapping the spatial patterns of internal migration in Europe. *Regional Studies, Regional Science* 7: 390–393.

Roy E. (2015) Welcome to Richie McCaw country – the town that made an All Blacks legend. *The Guardian*. Available at: https://www.theguardian.com/world/2015/oct/30/welcome-richie-mccaw-country-town-kurow-new-zealand-all-blacks.

Ruddick S. (1996) Constructing difference in public spaces: Race, class, and gender as interlocking systems. *Urban Geography* 17: 132–151.

Runte A. (1997) *National Parks: The American Experience*. Lincoln, NE: University of Nebraska Press.

Rye J and Gunnerud Berg N. (2011) The second home phenomenon and Norwegian rurality. *Norsk Geografisk Tidsskrift – Norwegian Journal of Geography* 65: 126–136.

Rye J and Slettebak M. (2020) The new geography of labour migration: EU11 migrants in rural Norway. *Journal of Rural Studies* 75: 125–131.

Sadhukhan J, Dugmore T, Matharu A, Martinez-Herandez E, Aburto J, Rahman P and Lynch J (2020) Perspectives on "game changer" global challenges for sustainable 21st century: Plant-based diet, unavoidable food waste biorefining, and circular economy. *Sustainability* 12: 1976.

Saint Onge J, Hunter L and Boardman J. (2007) Population growth in high-amenity rural areas: Does it bring socioeconomic benefits for long-term residents? *Social Science Quarterly* 88: 366–381.

Salvia R and Quaranta G. (2017) Place-based rural development and resilience: A lesson from a small community. *Sustainability* 9: 889.

Saint Andrews (2023) St Andrews Community https://www.standrewscommunity.ca/ Accessed 09/02/2023

Sands P and Peel J. (2012) *Principles of International Environmental Law*. Cambridge: Cambridge University Press.

Satsangi M. (2005) Landowners and the structure of affordable housing provision in rural Scotland. *Journal of Rural Studies* 21: 349–358.

Satsangi M. (2007) Land tenure change and rural housing in Scotland. *Scottish Geographical Journal* 123: 33–47.

Satsangi M and Gallent N. (2010) *The Rural Housing Question: Community and Planning in Britain's Countrysides*. Bristol: Policy Press.

Sauer C. (1925) The morphology of landscape. *University of California Publications in Geography* 2: 19–54.

Sauer C and Brand D. (1931) *Prehistoric Settlements of Sonora, with Special Reference to Cerros de Trincheras. University of California Publications in Geography 5.* Berkeley, CA: University of California Press.

Saunders P. (1984) Beyond housing classes: The sociological significance of private property rights in means of consumption†. *International Journal of Urban and Regional Research* 8: 202–227.

Scally C, Gilbert B, Hedman C, Gold A and Lily P. (2018) *Rental Housing for a 21st-Century Rural America: A Platform for Production.* Metropolitan Housing and Communities Policy Center.

Scally C and Tighe J. (2015) Democracy in action? NIMBY as impediment to equitable affordable housing siting. *Housing Studies* 30: 749–769.

Schama S. (1995) *Landscape and Memory.* Bath: Harper Collins.

Scott M. (2004) Building institutional capacity in rural Northern Ireland: The role of partnership governance in the LEADER II programme. *Journal of Rural Studies* 20: 49–59.

Scott M. (2013) Resilience: A conceptual lens for rural studies? *Geography Compass* 7: 597–610.

Scott M, Gallent N and Gkartzios M. (2019) *The Routledge Companion to Rural Planning.* London: Routledge.

Scott M and Murray M. (2009) Housing rural communities: Connecting rural dwellings to rural development in Ireland. *Housing Studies* 24: 755–774.

Sellick J. (2020) An introduction to new animal geographies: The case of cattle. *Geography* 105: 18–25.

Sellick J and Yarwood R. (2013) Placing livestock in landscape studies: Pastures new or out to graze? *Landscape Research* 38: 404–420.

Shaw H. (2014) *The Consuming Geographies of Food: Diet, Food Deserts and Obesity.* London: Routledge.

Shaw J. (1979) *Rural Deprivation and Planning.* Norwich: Geobooks.

Sheller M and Urry J. (2006) The new mobilities paradigm. *Environment and Planning A: Economy and Space* 38: 207–226.

Shergold I and Parkhurst G. (2012) Transport-related social exclusion amongst older people in rural Southwest England and Wales. *Journal of Rural Studies* 28: 412–421.

Sherman J. (2006) Coping with rural poverty: Economic survival and moral capital in rural America. *Social Forces* 85: 891–913.

Sherval M. (2009) Native Alaskan engagement with social constructions of rurality. *Journal of Rural Studies* 25: 425–434.

Shoard M. (1980) *The Theft of the Countryside.* Philadelphia: Temple Smith.

Short B. (1992) *The English Rural Community: Image and Analysis.* Cambridge: CUP Archive.

Short B. (2006) Idyllic ruralities. In: Cloke P, Goodwin M and Mooney P (eds) *Handbook of Rural Studies.* London: SAGE, 133–148.

Short J. (1991) *Imagined Country: Environment, Culture, and Society.* London; New York: Routledge.

Shortall O, Sutherland L, Ruston A, and Kaler J. (2018) True cowmen and commercial farmers: Exploring vets' and dairy farmers' contrasting views of 'good farming' in relation to biosecurity. *Sociologia Ruralis* 58: 583–603.

Shortall S. (2008) Are rural development programmes socially inclusive? Social inclusion, civic engagement, participation, and social capital: Exploring the differences. *Journal of Rural Studies* 24: 450–457.

Shortall S. (2016) Changing configurations of gender and rural society: Future directions for research. In: Shuksmith M and Brown D (eds) *Routledge International Handbook of Rural Studies.* London: Routledge, 425–433.

Shucksmith M. (1981) *No Homes for Locals?* Aldershot: Gower Publishing Company.

Shucksmith M. (1990) A theoretical perspective on rural housing. Housing classes in rural Britain. *Sociologia Ruralis* 30: 210–229.

Shucksmith M. (1991) Still no homes for locals? Affordable housing and planning controls in rural areas. In: Champion T and Watkins C (eds) *People in the Countryside*. London: Paul Chapman Publishing, 53–66.

Shucksmith M. (2000) Endogenous development, social capital and social inclusion: Perspectives from LEADER in the UK. *Sociologia Ruralis* 40: 208–218.

Shucksmith M. (2016) Social exclusion in rural places. In: Shucksmith M and Brown D (eds) *Routledge International Handbook of Rural Studies*. London: Routledge, 433–449.

Shucksmith M. (2018) Re-imagining the rural: From rural idyll to Good Countryside. *Journal of Rural Studies* 59: 163–172.

Shucksmith M and Brown D. (2018) Framing rural studies in the global north. In: Shucksmith M and Brown D (eds) *Routledge International Handbook of Rural Studies*. London: Routledge, 1–26.

Sica C. (2015) Stacked scale frames: Building hegemony for fracking across scales. *Area* 47: 443–450.

Sims R. (2010) Putting place on the menu: The negotiation of locality in UK food tourism, from production to consumption. *Journal of Rural Studies* 26: 105–115.

Singh A and Anitasingh. (2011) *Rihanna: Get off My Land and Find God, Farmer Tells Singer.* Available at: https://www.telegraph.co.uk/culture/8792072/Rihanna-get-off-my-land-and-find-God-farmer-tells-singer.html.

Skeffington Committee. (1969 [2013]) *People and Planning: Report of the Committee on Public Participation in Planning (The Skeffington Committee Report)*. London: Routledge.

Slominski S. (2020) *An Audience with Stephen Gough*. Available at: www.nakedrambler.org/an_audience_with_stephen_gough_by_stephen-slominski.html.

Smith D and Holt L. (2005) 'Lesbian migrants in the gentrified valley' and 'other' geographies of rural gentrification. *Journal of Rural Studies* 21: 313–322.

Smith K. (2019) Food system failure: Can we avert a future crisis? In: Shucksmith M and Brown D (eds) *Routledge International Handbook of Rural Studies*. London: Routledge, 250–261.

Smith L. (2013) Geographies of environmental restoration: A human geography critique of restored nature. *Transactions of the Institute of British Geographers* 38: 354–358.

Smith N. (1979) Toward a theory of gentrification a back to the city movement by capital, not people. *Journal of the American Planning Association* 45: 538–548.

Smith N, Davis A and Hirsch D. (2010) *A Minimum Income Standard for Rural Households*. York: JRF and Commission for Rural Communities.

Smith P, Bustamante H, Ahammad H, et al. (2014) Chapter 11 – Agriculture, forestry and other land use (AFOLU). In: Edenhofer O, Pichs-Madruga R, Sokona Y, Farahani E, Kadner K and Seyboth K. (eds) *Climate Change 2014: Mitigation of Climate Change. Contribution of Working Group III to the Fifth Assessment Report of the Intergovernmental Panel on Climate Change*. Cambridge: Cambridge University Press.

Solot M. (1986) Carl Sauer and cultural evolution. *Annals of the Association of American Geographers* 76: 508–520.

Soule M and Noss R. (1998) Rewilding and biodiversity: Complementary goals for continental conservation. *Wild Earth* 8: 18–28.

Spanier J. (2021) Rural futurism. *ACME: An International Journal for Critical Geographies* 20: 120–141.

Spence M. (1999) *Dispossessing the Wilderness: Indian Removal and the Making of the National Parks*. Oxford: Oxford University Press.

Squire S. (1993) Valuing countryside: Reflections on Beatrix Potter tourism. *Area* 5–10.

Staeheli L. (2008) Citizenship and the problem of community. *Political Geography* 27: 5–21.

Staeheli L. (2011) Political geography: Where's citizenship? *Progress in Human Geography* 35: 393–400.

Statistics Sweden. (2019) *Localities and Urban Areas 2018*. Available at: www.scb.se/en/finding-statistics/statistics-by-subject-area/environment/land-use/localities-and-urban-areas/.

Stenbacka S. (2011) Othering the rural: About the construction of rural masculinities and the unspoken urban hegemonic ideal in Swedish media. *Journal of Rural Studies* 27: 235–244.

Stevens S. (2014) *Indigenous Peoples, National Parks, and Protected Areas: A New Paradigm Linking Conservation, Culture, and Rights*. Tucson, AZ: University of Arizona Press.

Stillington and Whitton (2023) Home. https://www.stillington-whitton.com/Accessed 09/02/2023

Stockdale A. (2004) Rural out-migration: Community consequences and individual migrant experiences. *Sociologia Ruralis* 44: 167–194.

Stockdale A. (2006) Migration: Pre-requisite for rural economic regeneration? *Journal of Rural Studies* 22: 354–366.

Stockdale A. (2010) The diverse geographies of rural gentrification in Scotland. *Journal of Rural Studies* 26: 31–40.

Stockdale A. (2014) Unravelling the migration decision-making process: English early retirees moving to rural mid-Wales. *Journal of Rural Studies* 34: 161–171.

Stockdale A and Ferguson S. (2020) Planning to stay in the countryside: The insider-advantages of young adults from farm families. *Journal of Rural Studies* 78: 364–371.

Stockdale A, Findlay A and Short D. (2000) The repopulation of rural Scotland: Opportunity and threat. *Journal of Rural Studies* 16: 243–257.

Stoker, G. (1998). Public-Private Partnerships and Urban Governance. In: Pierre, J. (eds) Partnerships in Urban Governance. Palgrave Macmillan, London, pp.34-51

Storey D. (1999) Issues of integration, participation and empowerment in rural development: The case of LEADER in the Republic of Ireland. *Journal of Rural Studies* 15: 307–315.

Storey D. (2012) *Territories*. Milton Park; Abingdon; Oxon; New York: Routledge.

Storey D. (2013) 'New' migrants in the British countryside. *Journal of Rural and Community Development* 8: 291–302.

Storey D. (2016) Using heritage to promote rural places. *Town & Country Planning* 85: 286–288.

Strachan P and Lal D. (2004) Wind energy policy, planning and management practice in the UK: Hot air or a gathering storm? *Regional Studies* 38: 549–569.

Subramaniam Y, Masron T and Azman N. (2019) The impact of biofuels on food security. *International Economics* 160: 72–83.

Summers G. (1974) *Industrial Invasion of Nonmetropolitan America. A Quarter Century of Experience*. New York: Praeger.

Swaffield S and Fairweather J. (1998) In search of arcadia: The persistence of the rural idyll in New Zealand rural subdivisions. *Journal of Environmental Planning and Management* 41: 111–128.

Sylvia K, Peuch P and Franco J. (2015) *Extent of Farmland Grabbing in the EU*. Brussels: European Union.

Szekely V and Michniak D. (2009) Rural municipalities of Slovakia with a positive commuting balance. In: Neuwirth J and Wagner K (eds) *Multifunctional Territories: Importance of Rural Areas beyond Food Production*. Warsaw: Institute of Agricultural Economics and Food Economics, National Research Institute and the Federal Institute of Agricultural Economics.

Talanow K, Topp E, Loos J, Martin-Lopez B. (2021) Farmers' perceptions of climate change and adaptation strategies in South Africa's Western Cape. *Journal of Rural Studies* 81: 203–219.

Taylor M. (2008) *Living Working Countryside: The Taylor Review of Rural Economy and Affordable Housing*. London: Department for Communities and Local Government.

Taylor M and Walker C. (2015) *Garden Villages: Empowering Localism to Solve the Housing Crisis*. London: Policy Exchange.

Terluin I. (2003) Differences in economic development in rural regions of advanced countries: An overview and critical analysis of theories. *Journal of Rural Studies* 19: 327–344.

The Guardian. Escapist dreams: Why Germans love TV romances set in Cornwall. Available at: https://www.theguardian.com/world/2021/may/29/german-tv-love-cornwall-diplomats-rosamunde-pilcher.

The World Bank. (2018) *Piecing Together the Poverty Puzzle.* Washington, DC: The World Bank.

The World Bank (2023) *Employment in Agriculture.* https://data.worldbank.org/indicator/SL.AGR.EMPL.ZS?contextual=employment-by-sector&end=20 19&start=1991&view=chart Accessed 13/02/2023

Thomas N, Harvey D and Hawkins H. (2013) Crafting the region: Creative industries and practices of regional space. *Regional Studies* 47: 75–88.

Thorne C. (2014) Geographies of UK flooding in 2013/4. *The Geographical Journal* 180: 297–309.

Thrift N. (1987) Manufacturing rural geography? *Journal of Rural Studies* 3: 77–81.

Thrift N. (2008) *Non-representational Theory: Space, Politics, Affect.* London: Routledge.

Thuesen A. (2016) Gender and rural governance. In: Shucksmith M and Brown D (eds) *Routledge International Handbook of Rural Studies.* London: Routledge, 459–470.

Tickamyer A. (2006) Rural poverty. In: Cloke P, Marsden T and Mooney P (eds) *Handbook of Rural Studies.* London: SAGE, 411–426.

Tickamyer A, Sherman J and Warlick J. (2017) *Rural Poverty in the United States.* New York: Columbia University Press.

Todd J. (2021) Exploring trans people's lives in Britain, trans studies, geography and beyond: A review of research progress. *Geography Compass* 15: E12556.

Tolia-Kelly D. (2007) Fear in paradise: The affective registers of the English Lake District landscape re-visited. *The Senses and Society* 2: 329–351.

Tonnies F. (1955) *Community and Association.* London: Routledge.

Tonts M. (2001) The exclusive Brethren and an Australian rural community. *Journal of Rural Studies* 17: 309–322.

Tonts M. (2005) Competitive sport and social capital in rural Australia. *Journal of Rural Studies* 21: 137–149.

Tonts M and Atherley K. (2005) Rural restructuring and the changing geography of competitive sport. *Australian Geographer* 36: 125–144.

Tonts M and Larsen A. (2002) Rural disadvantage in Australia: A human rights perspective. *Geography* 87: 132–141.

Tonts M, Plummer P and Lawrie M. (2012) Socio economic wellbeing in Australian mining towns: A comparative analysis. *Journal of Rural Studies* 28: 288–301.

Tonts M, Yarwood R and Jones R. (2010) Global geographies of innovation diffusion: The case of the Australian cattle industry. *Geographical Journal* 176: 90–104.

Townsend P. (1987) Deprivation. *Journal of Social Policy* 16: 125–146.

Tuan Y. (1984) *Dominance & Affection: The Making of Pets.* New Haven, CT: Yale University Press.

Tuomisto H, Hodge I, Riordan P, and Macdonald D. (2012) Does organic farming reduce environmental impacts? – a meta-analysis of European research. *Journal of Environmental Management* 112: 309–320.

Tuomisto H and Teixeira de Mattos M. (2011) Environmental impacts of cultured meat production. *Environmental Science & Technology* 45: 6117–6123.

UNESCO. (2022) *The English Lake District.* Available at: https://whc.unesco.org/en/list/422.

United Nations. (1987) *Our Common Future: Report of the World Commission on Environment and Development.* New York: United Nations.

United Nations. (1995) *Programme of Action of the World Summit for Social Development.* New York: United Nations.

United Nations. (2020a) *2018 Revision of World Urbanization Prospects.* New York: United Nations.

United Nations. (2020b) *End Poverty in All Its Forms Everywhere.* Available at: https://unstats.un.org/sdgs/report/2019/goal-01/#:~:text=About%2079%20per%20cent%20of,under%2014%20years%20of%20age.

United Nations. (2022) *The 17 Goals.* Available at: https://sdgs.un.org/goals.

United States Census Bureau. (2020) *How Does the U.S. Census Bureau Define "Rural?".* Available at: https://gis-portal.data.census.gov/arcgis/apps/MapSeries/index.html?appid=7a41374f

6b03456e9d138cb014711e01#:~:text=The%20Census%20Bureau%20defines%20rural,tied%20to%20the%20urban%20definition.

United States Department of Agriculture. (2019) *Rural America at a Glance*. Washington, DC: United States Department of Agriculture.

Urbanik J. (2012) *Placing Animals: An Introduction to the Geography of Human-Animal Relations*. Lanham, MD: Rowman & Littlefield.

Urry J. (1990) *The Tourist Gaze: Leisure and Travel in Contemporary Societies*. London; Newbury Park, CA: SAGE.

Urry J. (1995) *Consuming Places*. London; New York: Routledge.

Urry J and Larsen J. (2011) *The Tourist Gaze 3.0*. Los Angeles, CA; London: SAGE.

Valentine G. (1996) Angels and devils: Moral landscapes of childhood. *Environment and Planning D: Society and Space* 14: 581–599.

Valentine G. (1997a) Making space: Lesbian separatist communities in the United States. In: Cloke P and Little J (eds) *Contested Countryside Cultures: Otherness, Marginalisation and Rurality*. London: Routledge, 109–122.

Valentine G. (1997b) A safe place to grow up? Parenting, perceptions of children's safety and the rural idyll. *Journal of Rural Studies* 13: 137–148.

Valentine G. (2001) *Social Geographies: Society and Space*. Harlow: Prentice Hall.

Valentine G, Holloway S, Knell C, and Jayne M. (2008) Drinking places: Young people and cultures of alcohol consumption in rural environments. *Journal of Rural Studies* 24: 28–40.

Vanderbeck R. (2003) Youth, racism, and place in the Tony Martin affair. *Antipode* 35: 363–384.

van der Ploeg J. (2000) Revitalizing agriculture: farming economically as starting ground for rural development. *Sociologia ruralis* 40: 497–511.

van der Ploeg J, Franco J and Borras S. (2015) Land concentration and land grabbing in Europe: A preliminary analysis. *Canadian Journal of Development Studies/Revue canadienne d'études du développement* 36: 147–162.

van Keulen J and Krijnen T. (2014) The limitations of localization: A cross-cultural comparative study of farmer wants a wife. *International Journal of Cultural Studies* 17: 277–292.

Velicu I and Kaika M. (2017) Undoing environmental justice: Re-imagining equality in the Rosia Montana anti-mining movement. *Geoforum* 84: 305–315.

Vepsäläinen M and Pitkänen K. (2010) Second home countryside. Representations of the rural in Finnish popular discourses. *Journal of Rural Studies* 26: 194–204.

Vera-Toscano E, Shucksmith M and Brown D. (2020) Poverty dynamics in Rural Britain 1991–2008: Did labour's social policy reforms make a difference? *Journal of Rural Studies* 75: 216–228.

Vergunst J. (2016) Changing environmental values. In: Shucksmith M and Brown D (eds) *Routledge International Handbook of Rural Studies*. London: Routledge, 285–294.

Vesalon L and Creţan R. (2015) 'We are not the Wild West': Anti-fracking protests in Romania. *Environmental Politics* 24: 288–307.

Vinet F. (2008) Geographical analysis of damage due to flash floods in southern France: The cases of 12–13 November 1999 and 8–9 September 2002. *Applied Geography* 28: 323–336.

Visser O, Sippel S and Thiemann L. (2021) Imprecision farming? Examining the (in) accuracy and risks of digital agriculture. *Journal of Rural Studies* 86: 623–632.

Visser O and Spoor M. (2011) Land grabbing in post-Soviet Eurasia: The world's largest agricultural land reserves at stake. *The Journal of Peasant Studies* 38: 299–323.

von Benzon N. (2017) Unruly children in unbounded spaces: School-based nature experiences for urban learning disabled young people in Greater Manchester, UK. *Journal of Rural Studies* 51: 240–250.

Von Thünen J. (1966) *Isolated State: An English Edition of Der Isolierte Staat*. Oxford: Pergamon Press.

Wackernagel M and Rees W. (1997) Perceptual and structural barriers to investing in natural capital: Economics from an ecological footprint perspective. *Ecological Economics* 20: 3–24.

Waegemakers Schiff J, Schiff R and Turner A. (2016) Rural homelessness in Western Canada: Lessons learned from diverse communities. *Social Inclusion* 4: 73–85.

Wainwright M. (2010) Britain's best views: The British Camp, Malvern Hills. *The Guardian*. Available at: https://www.theguardian.com/travel/2010/may/06/britains-best-views-worcester-malvern

Walford N. (2003) Productivism is allegedly dead, long live productivism. Evidence of continued productivist attitudes and decision-making in South-East England. *Journal of Rural Studies* 19: 491–502.

Walker H, Culham A, Fletcher A, and Reed M. (2019) Social dimensions of climate hazards in rural communities of the global North: An intersectionality framework. *Journal of Rural Studies* 72: 1–10.

Walker J and Cooper M. (2011) Genealogies of resilience: From systems ecology to the political economy of crisis adaptation. *Security Dialogue* 42: 143–160.

Walker M and Clark G. (2010) Parental choice and the rural primary school: Lifestyle, locality and loyalty. *Journal of Rural Studies* 26: 241–249.

Walmsley D, Epps W and Duncan C. (1998) Migration to the New South Wales north coast 1986–1991: Lifestyle motivated counterurbanisation. *Geoforum* 29: 105–118.

Walton J. (1984) The diffusion of the improved shorthorn breed of cattle in Britain during the eighteenth and nineteenth centuries. *Transactions of the Institute of British Geographers* 9: 22–36.

Wang W, Zhao X, Li H, and Zhang Q. (2021) Will social capital affect farmers' choices of climate change adaptation strategies? Evidences from rural households in the Qinghai-Tibetan Plateau, China. *Journal of Rural Studies* 83: 127–137.

Ward C. (1978) *The Child in the City*. New York: Pantheon Books.

Ward C. (1988) *The Child in the Country*. London: Bedford Square Press.

Ward N and McNicholas K. (1998a) Objective 5b of the structural funds and rural development in Britain. *Regional Studies* 32: 369–374.

Ward N and McNicholas K. (1998b) Reconfiguring rural development in the UK: Objective 5b and the new rural governance. *Journal of Rural Studies* 14: 27–39.

Watts D, Ilbery B and Maye D. (2005) Making reconnections in agro-food geography: Alternative systems of food provision. *Progress in Human Geography* 29: 22–40.

Weber C and Matthews H. (2008) Food-miles and the relative climate impacts of food choices in the United States. *Environmental Science & Technology* 42: 3508–3513.

Weber J and Sultana S. (2013) Why do so few minority people visit National Parks? Visitation and the accessibility of "America's Best Idea". *Annals of the Association of American Geographers* 103: 437–464.

Weekley I. (1988) Rural depopulation and counterurbanisation: A paradox. *Area* 127–134.

Weis T. (2013) *The Ecological Hoofprint: The Global Burden of Industrial Livestock*. London: Zed Books.

Wendt S. (2009) Constructions of local culture and impacts on domestic violence in an Australian rural community. *Journal of Rural Studies* 25: 175–184.

Westaway J. (2009) A sustainable future for geography? *Geography* 94: 4–12.

Westlund H, Larsson J and Olsson A. (2014) Start-ups and local entrepreneurial social capital in the municipalities of Sweden. *Regional Studies* 48: 974–994.

Whatmore S. (1988) From women's roles to gender relations – developing perspectives in the analysis of farm women. *Sociologia Ruralis* 28: 239–247.

Whatmore S. (1991a) *Farming Women: Gender, Work, and Family Enterprise*. Houndmills; Basingstoke; Hampshire: Macmillan Academic and Professional.

Whatmore S. (1991b) Life-cycle or patriarchy – gender divisions in family farming. *Journal of Rural Studies* 7: 71–76.

Whatmore S. (1997) Dissecting the autonomous self: Hybrid cartographies for a relational ethics. *Environment and Planning D: Society and Space* 15: 37–53.

Whatmore S. (1999) Rethinking the "human" in human geography. In: Massey D, Sarre P and Allen J (eds) *Human Geography Today*. New York: Wiley, 22–41.

Whatmore S. (2002) *Hybrid Geographies: Natures, Cultures, Spaces*. London; Thousand Oaks, CA: SAGE.

Wheeler R. (2014) Mining memories in a rural community: Landscape, temporality and place identity. *Journal of Rural Studies* 36: 22–32.

Wheeler R. (2017) Reconciling Windfarms with rural place identity: Exploring residents' attitudes to existing sites. *Sociologia Ruralis* 57: 110–132.

Whyte I. (2000) William Wordsworth's guide to the lakes and the geographical tradition. *Area* 32: 101–106.

Wilkie R. (2005) Sentient commodities and productive paradoxes: The ambiguous nature of human – livestock relations in Northeast Scotland. *Journal of Rural Studies* 21: 213–230.

Williams R. (1973) *The Country and the City*. New York: Oxford University Press.

Willis R. (2004) Enlargement, concentration and centralisation in the New Zealand dairy industry. *Geography* 89: 83–88.

Wilson G. (2001) From productivism to post-productivism . . . and back again? Exploring the (un) changed natural and mental landscapes of European agriculture. *Transactions of the Institute of British Geographers* 26: 77–102.

Wilson G. (2010) Multifunctional 'quality' and rural community resilience. *Transactions of the Institute of British Geographers* 35: 364–381.

Wilson G. (2012) *Community Resilience and Environmental Transitions*. London: Routledge.

Wilson G and Burton R. (2015) 'Neo-productivist' agriculture: Spatio-temporal versus structuralist perspectives. *Journal of Rural Studies* 38: 52–64.

Wilson G and Dyke S. (2016) Pre- and post-installation community perceptions of wind farm projects: The case of Roskrow Barton (Cornwall, UK). *Land Use Policy* 52: 287–296.

Wilson G and Rigg J. (2003) 'Post-productivist' agricultural regimes and the South: Discordant concepts? *Progress in Human Geography* 27: 681–707.

Wilson O and Klages B. (2001) Farm restructuring in the ex-GDR: Towards a new farm model? *Journal of Rural Studies* 17: 277–291.

Winchester H, Kong H and Dunn K. (2013) *Landscapes: Ways of Imagining the World*. London: Routledge.

Winter M. (1996) *Rural Politics: Policies for Agriculture, Forestry and the Environment*. London: Psychology Press.

Winter M. (2003) Embeddedness, the new food economy and defensive localism. *Journal of Rural Studies* 19: 23–32.

Winter M. (2006) Rescaling rurality: Multilevel governance of the agro-food sector. *Political Geography* 25: 735–751.

Witten K, Kearns R, Lewis N, Coster H and McCreanor T. (2003) Educational restructuring from a community viewpoint: A case study of school closure from Invercargill, New Zealand. *Environment and Planning C: Government and Policy* 21: 203–223.

Wolch J. (2002) Anima Urbis. *Progress in Human Geography* 26: 721–742.

Wolch J and Emel J. (1995) Guest editorial: Bringing the animals back in. *Environment and Planning D: Society and Space* 13: 632–636.

Wolch J and Emel J. (1998) *Animal Geographies: Place, Politics and Identity in the Nature-Culture Borderlands*. London: Verso.

Women and Geography Study Group. (1984) *Geography and Gender: An Introduction*. London: Hutchison.

Woods M. (1998) Researching rural conflicts: Hunting, local politics and actor-networks. *Journal of Rural Studies* 14: 321–340.

Woods M. (2003) Conflicting environmental visions of the rural: Windfarm development in Mid Wales. *Sociologia Ruralis* 43: 271–288.

Woods M. (2003) Deconstructing rural protest: the emergence of a new social movement. *Journal of Rural Studies* 19: 309–325.

Woods M. (2004) *Rural Geography: Processes, Responses and Experiences in Rural Restructuring*. London: SAGE.

Woods M. (2006) Redefining the 'rural question': The new 'politics of the rural' and social policy. *Social Policy & Administration* 40: 579–595.

Woods M. (2007) Engaging the global countryside: Globalization, hybridity and the reconstitution of rural place. *Progress in Human Geography* 31: 485–507.

Woods M. (2010) Performing rurality and practising rural geography. *Progress in Human Geography* 34: 835–846.

Woods M. (2011a) The local politics of the global countryside: Boosterism, aspirational ruralism and the contested reconstitution of Queenstown, New Zealand. *GeoJournal* 76: 365–381.

Woods M. (2011b) *Rural*. London: Routledge.

Woods M. (2012) Rural geography III: Rural futures and the future of rural geography. *Progress in Human Geography* 36: 125–134.

Woods M. (2016) Policing rural protest. In: Mawby R and Yarwood R (eds) *Rural Policing and Policing the Rural*. London: Routledge, 123–136.

Woods M. (2019) The future of rural places. In: Scott M, Gallent N and Gkartios M (eds) *The Routledge Companion to Rural Planning*. London: Routledge, 622–632.

Woods M, Fois F, Heley J, Jones L, Onyeahialam A, Saville S and Welsh M. (2021) Assemblage, place and globalisation. *Transactions of the Institute of British Geographers* 46(2): 284–298.

Woodward R. (1996) 'Deprivation' and 'the rural': An investigation into contradictory discourses. *Journal of Rural Studies* 12: 55–67.

Woodward R. (2000) Warrior heroes and little green men: Soldiers, military training, and construction of rural masculinities. *Rural Sociology* 65: 640–657.

Woodward R. (2004) *Military Geographies*. London: John Wiley & Sons.

Wooff A. (2015) Relationships and responses: Policing anti-social behaviour in rural Scotland. *Journal of Rural Studies* 39: 287–295.

Woolcock M. (1998) Social capital and economic development: Toward a theoretical synthesis and policy framework. *Theory and Society* 27: 151–208.

World Health Organisation. (2020) *Social Determinants of Health*. Available at: www.who.int/social_determinants/themes/socialexclusion/en/.

Wylie J. (2005) A single day's walking: Narrating self and landscape on the South West Coast path. *Transactions of the Institute of British Geographers* 30: 234–247.

Wylie J. (2007) *Landscape*. London: Routledge.

Wylie J. (2009a) Landscape. In: Gregory D, Johnston R, Pratt G, Watts M and Whatmore S. (eds) *The Dictionary of Human Geography*. Oxford: Wiley-Blackwell, 409–411.

Wylie J. (2009b) Landscape, absence and the geographies of love. *Transactions of the Institute of British Geographers* 34: 275–289.

Yair K and Schwarz M. (2011) Making value: Craft in changing times. *Cultural Trends* 20: 309–316.

Yarwood R. (1996) Rurality, locality and industrial change: A micro-scale investigation of manufacturing growth in the district of Leominster. *Geoforum* 27: 23–37.

Yarwood R. (2002a) *Countryside Conflicts*. Sheffield: Geographical Association.

Yarwood R. (2002b) Parish councils, partnership and governance: The development of 'exceptions' housing in the Malvern Hills District, England. *Journal of Rural Studies* 18: 275–291.

Yarwood R. (2005a) Beyond the rural idyll – Images, countryside change and geography. *Geography* 90: 19–31.

Yarwood R. (2005b) Crime concern and policing in the countryside: Evidence from Parish councillors in West Mercia Constabulary, England. *Policing and Society* 15: 63–82.

Yarwood R. (2007a) The geographies of policing. *Progress in Human Geography* 31: 447–465.

Yarwood R. (2007b) Getting just deserts? Policing, governance and rurality in Western Australia. *Geoforum* 38: 339–352.

Yarwood R. (2010) An exclusive countryside? Crime concern, social exclusion and community policing in two English villages. *Policing and Society* 20: 61–78.

Yarwood R. (2012) Neighbourhood watch. In: Smith S (ed.) *International Encyclopedia of Housing and Home*. Oxford: Elsevier, 90–95.

Yarwood R. (2012) One moor night: emergencies, training and rural space. *Area* 44: 22–28.

Yarwood R. (2014) *Citizenship*. London: Routledge.

Yarwood R. (2015) Lost and hound: The more-than-human networks of rural policing. *Journal of Rural Studies* 39: 278–286.

Yarwood R. (2017) Rural citizenship. In: *The International Encyclopedia of Geography: People, the Earth, Environment, and Technology*. New York: Wiley.

Yarwood R. (2020a) Citizenship, locality and territory. In: Storey D (ed.) *A Research Agenda for Territory and Territoriality*. Cheltenham: Edward Elgar, 145–158.

Yarwood R. (2020b) Territory, locality and citizenship. In: Storey D (ed.) *A Research Agenda for Territory and Territoriality*. Cheltenham: Edward Elgar Publishing.

Yarwood R. (2021) Policing the global countryside: Towards a research agenda. *Professional Geographer*, Forthcoming.

Yarwood, R (2023) Rethinking British National Parks: country, coasts and cities for all? *Geography* 108 (1), 6-16

Yarwood R and Absalom T. (2006) Devon livestock breeds: A geographical perspective. *Transactions of Devonshire Association for the Advancement of Science* 138: 93–130.

Yarwood R and Charlton C. (2009) 'Country life'? Rurality, folk music and 'Show of hands'. *Journal of Rural Studies* 25: 194–206.

Yarwood R and Edwards B. (1995) Voluntary action in rural areas: The case of neighbourhood watch. *Journal of Rural Studies* 11: 447–459.

Yarwood R and Evans N. (1999) The changing geography of rare livestock breeds in Britain. *Geography: Journal of the Geographical Association* 84: 80–87.

Yarwood R and Evans N. (2000) Taking stock of farm animals and rurality. In: Philo C and Wilbert C (eds) *Animals Spaces, Beastly Places*. London: Routledge, 98–114.

Yarwood R and Evans N. (2003) Livestock, locality and landscape: EU regulations and the new geography of Welsh farm animals. *Applied Geography* 23: 137–157.

Yarwood R and Evans N. (2006) A Lleyn sweep for local sheep? Breed societies and the geographies of Welsh livestock. *Environment and Planning A: Economy and Space* 38: 1307–1326.

Yarwood R and Gardner G. (2000) Fear of crime, cultural threat and the countryside. *Area* 32: 403–411.

Yarwood R and Shaw J. (2010) 'N-gauging' geographies: Craft consumption, indoor leisure and model railways. *Area* 42: 425–433.

Yarwood R, Sidaway J, Kelly C, Stillwell S. (2015) Sustainable deathstyles? The geography of green burials in Britain. *Geographical Journal* 181: 172–184.

Yarwood R, Tonts M and Jones R. (2010) The historical geographies of showing livestock: A case study of the Perth Royal Show, Western Australia. *Geographical Research* 48: 235–248.

Yashar D. (1998) Contesting citizenship: Indigenous movements and democracy in Latin America. *Comparative Politics* 23–42.

Yeo S. (2003) Flood risk management for caravan parks in New South Wales. *Australian Geographer* 34: 195–209.

Zoomers A. (2010) Globalisation and the foreignisation of space: Seven processes driving the current global land grab. *The Journal of Peasant Studies* 37: 429–447.

Index

Note: Page numbers in italics refer to figures, those in bold refer to tables.